Electrotechnical Systems

Calculation and Analysis
with Mathematica® and PSpice®

Electrotechnical Systems

Calculation and Analysis
with Mathematica® and PSpice®

Igor Korotyeyev

Valeri Zhuikov

Radoslaw Kasperek

CRC Press
Taylor & Francis Group
Boca Raton   London   New York

CRC Press is an imprint of the
Taylor & Francis Group, an **Informa** business

CRC Press
Taylor & Francis Group
6000 Broken Sound Parkway NW, Suite 300
Boca Raton, FL 33487-2742

First issued in paperback 2017

© 2010 by Taylor and Francis Group, LLC
CRC Press is an imprint of Taylor & Francis Group, an Informa business

No claim to original U.S. Government works

ISBN 13: 978-1-138-11775-4 (pbk)
ISBN 13: 978-1-4200-8709-3 (hbk)

Library of Congress Cataloging-in-Publication Data

Korotyeyev, Igor.
    Electrotechnical systems : calculation and analysis with Mathematica and PSpice / Igor Korotyeyev, Valeri Zhuikov, Radoslaw Kasperek.
        p. cm.
    "A CRC title."
    Includes bibliographical references and index.
    ISBN 978-1-4200-8709-3 (hardcover : alk. paper)
    1. Electronic systems--Design and construction--Data processing. 2. Electronic circuits--Design--Data processing. 3. Mathematica (Computer file) 4. PSpice. I. Zhuikov, Valeri. II. Kasperek, Radoslaw. III. Title.

TK7870.K6526 2010
621.381--dc22   2009046417

Visit the Taylor & Francis Web site at
http://www.taylorandfrancis.com

and the CRC Press Web site at
http://www.crcpress.com

Contents

Preface

The development of mathematical methods and analysis, and computer technology with advanced electrotechnical devices has led to the creation of various programs increasing labor productivity. There are three types of programs: mathematical, simulation, and programs that unite these two operations. Furthermore, these programs are often used for analysis in various areas.

Mathematical programs perform analytic and numerical methods and transformations that realize known mathematical operations. Among the better-known programs are Mathematica® and Maple®.

Programs that carry out the analysis of electromagnetic processes in electronic and electrotechnical devices and systems belong to the family of simulation programs. Such programs have additional abilities such as the calculation of thermal conditions, sensibility, and harmonic composition. One such widely known program is ORCAD (formerly PSpice®), which allows modeling of digital devices and the design of printed circuit cards. We are interested in programs in which the mathematical description and methods, together with methods of modeling, are incorporated in the general software product. The most widespread program is MATLAB®. MATLAB's potential is enhanced by the inclusion in its structure of various up-to-date methods, such as neural networks and systems of fuzzy logic.

The characteristics of the programs are presented here briefly, showing the relative niche occupied by each program. Depending on the problems in question (e.g., programmer qualification, capabilities of the program), we can effectively analyze enough complex systems. In some cases preference is given to mathematical programs that include a powerful block of analytic transformations. It is expedient to use a simulation program if it is necessary to develop and analyze electronic systems. There are certain limitations in their use caused by the elements involved in a program. Another deficiency is the absence of a maneuver, as in the analysis of stiff systems. In such a case, as a rule, it is necessary to change the model of the elements or change the purpose or the model of the whole system. For example, during the determination of a steady-state process, the system may be unstable. In this case, use of the simulation programs does not give the answer to the question of what is necessary to change in the system in order to maintain its working capacity. For this, it is necessary to undertake an additional analysis of the model. And in this case mathematical programs have an advantage in respect to the ability of formation and change of complexity of the model, and to a choice of mathematical methods used in the solution of a problem. This feature of mathematical programs is very attractive for researchers, and is the main reason why authors select the mathematical program as the tool for research.

The application of the mathematical pocket Mathematica 4.2 for the analysis of the electromagnetic processes in electrotechnical systems is shown in this book. For the clarity of represented expressions, and expressions, variables, and functions used by Mathematica for the input, the latter will be shown in bold.

MATLAB® is a registered trademark of The MathWorks, Inc. For product information, please contact:

The MathWorks, Inc.
3 Apple Hill Drive
Natick, MA 01760-2098 USA
Tel: 508 647 7000
Fax: 508-647-7001
E-mail: info@mathworks.com
Web: www.mathworks.com

Acknowledgments

I would like to give special thanks to Prof. Zbigniew Fedyczak with whom I have worked over the last few years on matrix reactance converters. I am also grateful to Kiev Polytechnic Institute for its teachers and instilling in me the rigors of a scientist. I cannot omit to acknowledge my thanks to the University of Zielona Gora, which has afforded me the opportunity to write this book.

My wife Lyudmila, my daughter Lilia, son-in-law Volodya, and my grandchildren Volodya and Kolya have been constant supports in my scientific work and the writing of this book. My parents have been a pillar of support in my efforts to solve intricate problems and have encouraged my perseverance in doing so.

Igor Korotyeyev

Many different factors have influenced the appearance of this work, not the least of which is the important and longstanding good relations between the University of Zielona Góra, Poland, and the National Technical University of Ukraine (Kiev Polytechnic Institute [KPI]). Such good relations have been at all times supported by many specialists, and in this respect I would like to emphasize my profound gratitude to Prof. Jozef Korbiez, Prof. Zbigniew Fedyczak, and Prof. Ryszard Strzelski (Gdynia Maritime University) who has done much for the development of our friendly relations. I am particularly grateful to Prof. Vladimir Rudenko, my adviser and teacher, and founder of the industrial electronics department of the KPI. I am aware that I have much to thank him for in my achievements, and for his contributions to my achievements that I am not aware of, I also thank him.

Valeri Zhuikov

It is with great humility that I acknowledge the guidance, support, and advice that I have received from my family, friends, and colleagues in their unselfish help, motivation, indulgence, and patience. I would like to express my appreciation to all those persons who have devoted their precious time to helping me in my work on this book.

Radosław Kasperek

Finally, the authors acknowledge the painstaking efforts of Peter Preston in the improvement of the language of our manuscript.

The Authors

Igor Korotyeyev was born in Kiev, Ukraine, in 1950. He received his diploma in engineering in industrial electronic from the Kiev Polytechnic Institute in 1973, and a Ph.D. degree and D.Tech.S. degree from the Institute of Electrodynamics, Kiev, in 1979 and 1994, respectively.

He was with Kiev Polytechnic Institute as an assistant professor from 1979 to 1995. Since 1995, he was appointed a full professor in industrial electronics at Kiev Polytechnic Institute, and since  1998, has taught industrial electronics at the University of Zielona Gora, Poland, where he is a full professor. His fields of interests are process modeling and stability investigation in power converters.

Valeri Zhuikov was born in 1945. He received his Ph.D. degree in 1975, and in 1986 he was awarded the Dr.Sc. degree. Now he is dean of the electronics faculty, the head of the Department of Industrial Electronics, National Technical University of Ukraine (Kiev Polytechnical Institute). His field of interest is the theory of processes estimation in power electronics systems.

Radosław Kasperek was born in 1970 in Zielona Góra, Poland. He received an M.Sc. degree in electrical engineering from the Technical University of Zielona Góra in 1995 and then joined the Institute of Electrical Engineering there. In 2004 he received a Ph.D. degree in electrical engineering from the Department of Electrical Engineering, Computer Science and Telecommunication, University of Zielona Góra. His fields of interests are electrical machines, power converters, and power quality.

1

Characteristics of the Mathematica® System

1.1 Calculations and Transformations of Equations

An elementary example of the use of Mathematica® is the execution of calculations with the sphere of the calculator. Let us input the following expression to the Mathematica notepad:

12/3

and then press the keys Shift + Enter. The expression In[1] = will appear to the left of this expression, and in the next row,

Out[2] = 4

As we have entered integer numbers, Mathematica has calculated the result as an integer value. For the expression

11/3

Mathematica displays

$$\frac{11}{3}$$

Let us use the built-in function **N[ ]** of Mathematica. Then, for

N[11/3]

we get

3.66667

Built-in functions of Mathematica begin with the capital letters, and the argument is enclosed in square brackets.

There is an alternative calculation. For this purpose, at the end of equation, it is necessary to write down **//N**, that is,

$$11/3//N$$

When real numbers are entered, Mathematica executes the calculation without the use of function **N[ ]**. For example, for

$$12.2/3$$

we have

$$4.06667$$

Real numbers are entered in the format

$$1.22*10\wedge1$$

$$122.0*10\wedge{-}1$$

The multiplier sign is entered either by the space or by the asterisk; the degree sign is entered with the help of the symbol ∧.

Complex numbers are inputted with the help of the symbol of imaginary unit **I** (or **i**). For example,

$$1.2+I*3.2$$

Calculations with complex numbers are also executed just as with real ones. For example, for the result of the calculation

$$(1.2+I*3.2)/(2.0+I*9.1)$$

we obtain

$$0.363092{-}0.0520677i$$

Real and imaginary parts of complex numbers are distinguished with the help of the functions **Re[ ]** and **Im[ ]**. For example,

$$Re[6.1{-}I*5.5]$$

$$Im[6.1{-}I*5.5]$$

$$6.1$$

$$-5.5$$

In Mathematica, use of some constants for which symbols are reserved is provided: imaginary unit **I** (or **i**), **E** (the base of the natural logarithm), **Pi** (π number), **Degree** ($\pi/180$ number), and **Infinity** (infinity) are some of them.

When complex systems are calculated, names are given to the variables called *named variables*. A named variable begins with a letter. The value of the variable is assigned by means of an operation of assignment. For example, for

$$con1=56.2;$$

$$con2=14.7;$$

$$con1/con2$$

we have

$$3.82313$$

We write values of parameters in each row of the cell of a notepad. Several parameters can be entered in one row, but they must be separated by the semicolon sign (;). When the semicolon sign is not written at the end of the row, then the parameter value will be written down in a separate cell after the cell calculation. It is also necessary to keep in mind that a line feed is made by pressing the Enter key.

One more way of assigning the value of a variable is determined by the sign: =. For example,

$$var1:=var2;$$

In this case, the right part will not be calculated, while the variable **var1** will not appear in following expressions. Let us consider by examples the difference between the presented assignment techniques. In the first example,

$$con1=16.2;$$

$$con2=4;$$

$$var1=con1/con2$$

$$con2=3;$$

$$var1$$

we obtain

$$4.05$$

$$4.05$$

In the second example,

con1=16.2;

con2=4;

var1:=con1/con2;

var1

con2=3;

var1

we obtain

4.05

5.4

Thus, we can change the value of a variable during the calculations.

During calculations of various expressions, it is often necessary to carry out their transformations. The **Expand[ ]** function permits expansion of products. For example, calculating

var1=(x+3.9)*(y−2.1);

var2=Expand[var1]

yields

$$-8.19 - 2.1x + 3.9y + xy$$

We can transform the obtained expression for the given variable with the help of the function **Collect[ ]**. Applying

Collect[var2,x]

yields

$$-8.19 + x(-2.1 + y) + 3.9y$$

For the expansion of polynomials with integer numbers, the function **Factor[ ]** is used. Applying this function to the expression

var1=x*y+3*y-2*x-6;

Factor[var1]

yields

$$(3+x)(-2+y)$$

The function **Simplify[ ]** produces the algebraic manipulation of an argument and returns its simple form. If in the considered example we replace the function **Factor[ ]** with **Simplify[ ]**, the result will be the same. The functions **Simplify[ ]** and **Factor[ ]** in analytical transformations also allow us to effect reduction of fractions. For example, for

$$var1=x/(x+1)-2/(x^2-1);$$

$$Simplify[var1]$$

we obtain

$$\frac{-2+x}{-1+x}$$

In Mathematica, the function **FullSimplify[ ]**, in comparison with the function **Simplify[ ]**, has a greater range of capabilities. Let us show the difference between these two functions with the example:

$$var1=(x*y+4*x+3.1*y+12.4)/(x+3.1);$$

$$Simplify[var1]$$

$$FullSimplify[var1]$$

As a result of the use of the first function, we obtain

$$\frac{12.4+3.1y+x(4+y)}{3.1+x}$$

for the second

$$4.+y$$

For reduction of the common multipliers in the numerator and denominator, the **Cancel[ ]** function is used. The transformed expression must be represented in the form of a fraction. Then, for

$$Cancel[(s*d+a*s+h*d+a*h)/(s+h)]$$

we obtain

$$a+d$$

The **Together[ ]** function allows the reduction of fractions to the common denominator and the cancellation of the common multipliers in the numerator and denominator. For the expression

$$var1=x^2/(x-1)+(-2*x+1)/(x-1);$$

$$Together[var1]$$

we obtain

$$-1+x$$

It should be noted that, for this example, the application of the **Simplify[ ]** and **Factor[ ]** functions allow us to obtain the same result.

The **Apart[ ]** function presents an argument as a sum of fractions. As a result of the application of this function to the expression

$$var1=(x^2-2*x*y+y^2-x^2*y^2)/(x^2-2*x*y+y^2);$$

$$Apart[var1]$$

we obtain

$$1-x^2-\frac{x^4}{(-x+y)^2}-\frac{2x^3}{-x+y}$$

The substitutions are often used during the transformation of the expressions in Mathematica. A substitution operation is determined by the symbol /.. The expression following this symbol, **var1->var2**, shows that **var2** replaces the variable **var1**. The symbol -> consists of two symbols: - and >. Let us consider the example of the application of substitution

$$x=a+4;$$

$$m=x/.a->z+3;$$

$$y=b+6;$$

$$b=z+1;$$

$$y$$

$$m$$

$$x$$

As a result we obtain

$$7+z$$

$$7+z$$

$$4+a$$

Thus, the first equation remained unchangeable for **x**, but the equation for **y** changed.

1.2 Solutions of Algebraic and Differential Equations

The **Solve[ ]** function is used for solutions of algebraic equations. Let us find the solution to the algebraic equation

$$x^2 - 1.6x - 7.77 = 0$$

We shall define the variable corresponding to the equation and apply the **Solve[ ]** function:

eq1=x^2-1.6*x-7.77;

x12=Solve[eq1 == 0,x]

The first part of the **Solve[ ]** involves the equation (or system of equations), but the second part involves the variable (or list of variables), according to which the equation must be solved. The sign == is obtained by way of entering two signs of =. The result of the solution is represented as the list

$$\{\{x \rightarrow -2.1\}, \quad \{x \rightarrow 3.7\}\}$$

in which the substitutions are used. For assignment of the solution to the variables **x1** and **x2**, it is necessary to use the substitution of the solution **x12** for the variables and then pick out the separate values. Continuing the previous example,

x1=Part[x/.x12,1]

x2=Part[x/.x12,2]

we obtain

−2.1

3.7

By means of the **Part[ ]** function, extraction of the element from the list is made.

For the set of equations

$$eq1=a*x+b*y+c;$$

$$eq2=2*a*x+2*b*y+2*c;$$

$$xy=Solve[\{eq1 == 0, \quad eq2 == 0\},\{x,y\}]$$

Mathematica displays
   Solve::svars: Equations may not give solutions for all "solve" variables.

$$\left\{\left\{x \to -\frac{c}{a}-\frac{by}{a}\right\}\right\}$$

Change the second equation in the following way and apply the **Solve[ ]** function

$$eq1=a*x+b*y+c;$$

$$eq2=2*a*x+2*b*y+c;$$

$$xy=Solve[\{eq1 == 0,eq2 == 0\},\{x,y\}]$$

We obtain the answer

$$\{\}$$

which shows that there is no solution.
   Change the second equation once again. As a result of solving the set of equations

$$eq1=a*x+b*y+c;$$

$$eq2=2*a*x+b*y+c;$$

$$xy=Solve[\{eq1 == 0,eq2 == 0\},\{x,y\}]$$

we obtain

$$\left\{\left\{x \to 0, y \to -\frac{c}{b}\right\}\right\}$$

Use the **Part[ ]** function to assign the solution to the variables

$$x1=Part[x/.xy,1]$$

$$y1=Part[y/.xy,1]$$

Then

$$0$$

$$-\frac{c}{b}$$

For elimination of a part of the variables from the set of equations, it is necessary to use the **Eliminate[ ]** function. If we use the equations from the last example, then for

eq3=Eliminate[{eq1==0,eq2==0},x]

we obtain

$$-by == c$$

The solution to this equation can be found with the help of the **Solve[ ]** function.

For the numeral solution to the algebraic equations, the **NSolve[ ]** function is used. For example, for the equation

eq1=x^5-2*x^2 ι 3;

NSolve[eq1 == 0,x]

we obtain

$$\{\{x \to -1.\}, \quad \{x \to -0.585371 - 1.34012i\}, \quad \{x \to -0.585371 + 1.34012i\}\}$$

When equations are represented in the matrix form, it is expedient to use the **LinearSolve[ ]** function for their solution.

For the numeral solution to nonlinear equations in Mathematica, the **FindRoot[ ]** function is used. In this function, the initial value is introduced and, in case of need, the interval on which the solution will be found is also introduced. For example, solving the equation

$$e^{-x} = x$$

by means of

FindRoot[Exp[-x]==x,{x,1}]

yields

$$\{x \to 0.567143\}$$

The second argument {x,1} of the function in this case defines the initial value and the variable according to which the solution is calculated.

With the solving of the differential equations in Mathematica, it is necessary to set both a function and independent variable according to which the solution is found. We find the solution to the 2nd-order differential equation

$$\frac{d^2y}{dx^2} + 2\frac{dy}{dx} + 3y = 0.$$

Using the **DSolve[ ]** function

$$\text{eq1=y''[x]+2*y'[x]+3*y[x];}$$

$$\text{s1=DSolve[eq1 == 0,y[x],x]}$$

we obtain the solution

$$\left\{\left\{y[x] \to e^{-x}C[2]Cos[\sqrt{2}x] + e^{-x}C[1]Sin[\sqrt{2}x]\right\}\right\}$$

in which two constants C[1] and C[2] are presented. To extract the solution, the **Part[ ]** function is used

$$\text{ys=Part[y[x]/.s1,1]}$$

Then,

$$e^{-x}C[2]Cos[\sqrt{2}x] + e^{-x}C[1]Sin[\sqrt{2}x]$$

Let us calculate the value of this expression at the point $x = 2$ at C[1] = 3 and C[2] = 4

$$\text{X=2;}$$

$$\text{yd=ys/.\{C[1]->3,C[2]->4\}}$$

We obtain

$$\frac{4Cos[2\sqrt{2}]}{e^2} + \frac{3Sin[2\sqrt{2}]}{e^2}$$

The numerical value is determined with the help of the **N[ ]** function

$$\text{N[yd]}$$

Then, mathematica outputs

$$-0.389933$$

The **DSolve[ ]** function is used for the solution to the set of differential equations. We solve the set of the first-order differential equations

$$\frac{dy}{dt} - 3*y + x = 0,$$

$$\frac{dx}{dt} + 2*x - y = 1$$

with the initial conditions $y(0) = -1$, $x(0) = 2$. The set of equations is represented as follows:

eq1=y′[t]-3*y[t]+x[t];

eq2=x′[t]+2*x[t]-y[t]-1;

As a result of the solution

s1=DSolve[{eq1==0,eq2==0,y[0]==-1,x[0]==2},{y[t],x[t]},t]//N

we obtain

$$\{\{y[t] \rightarrow 0.0952381(21. + 37.8167 \cdot 2.71828^{-1.79129t} - 163.8117 \cdot 2.711828^{2.79129t}),$$

$$x[t] \rightarrow 0.0047619(126. + 362.381 \cdot 2.71828^{-1.79129t} - 68.3811 \cdot 2.711828^{2.79129t})\}\}$$

Remember that the **//N** function specifies that the solution should be obtained in a numeral form.

Let us transform this solution in the following way:

Simplify[s1]

Then,

$$\{\{y[t] \rightarrow 0.2 + 0.360159e^{-1.79129t} - 1.56016e^{2.79129t},$$

$$\{x[t] \rightarrow 0.6 + 1.72562e^{-1.79129t} - 0.325624e^{2.79129t}\}\}$$

For the numeral solution to differential equations in Mathematica, the function **NDSolve[ ]** is used. Let us find the solution to the same system on the interval 0 … 1.

eq1=y′[t]−3*y[t]+x[t];

eq2=x′[t]+2*x[t]-y[t]-1;

s2=NDSolve[{eq1==0,eq2==0,y[0]==-1,x[0]==2},{y,x},{t,0,1}]

As a result of the application of the function **NDSolve[ ]**, we obtain the solution in the form of interpolation functions

$$\{\{y\text{->}InterpolatingFunction[\{\{0.,1.\}\},\diamond],$$

$$x\text{->} InterpolatingFunction[\{\{0.,1.\}\},\diamond]\}\}$$

For $t = 0.2$, the value of functions is obtained in the following way:

$$Part[y[0.2]/.s2,1]$$

$$Part[x[0.2]/.s2,1]$$

Then

$$-2.27486$$

$$1.23696$$

1.3 Use of Vectors and Matrices

In Mathematica the vectors and matrices are represented in the view of lists. For example, vector $u = \{0.1, 0.25\}$, matrix $m = \{\{a, b\}, \{c, d\}\}$. There are various functions in Mathematica to work with vectors and matrices. Let us consider an example. We find the inverse matrix for

$$m1 = \begin{pmatrix} 0 & 0 \\ 0.1 & 0.2 \end{pmatrix};$$

Inverse[m1]

Mathematica displays:

Inverse::sing: Matrix$\{\{0.,0.\},\{0.1,0.2\}\}$ is singular

Inverse$[\{\{0,0\},\{0.1,0.2\}\}]$

Mathematica informs that the matrix is singular. Let us find the eigenvalues of the matrix with the help of the function

Eigenvalues[m1]

Then

$$\{0.2, 0.\}$$

In fact, one of the eigenvalues of the matrix is equal to zero.

Let us change the data of the example. Consider the matrix

$$\text{m1} \quad \begin{pmatrix} 0.3 & 2.0 \\ 0.1 & 0.2 \end{pmatrix};$$

Applying the function

Inverse[m1]

Eigenvalues[m1]

yields

$$\{\{0.769231, 7.69231\}, \{-0.384615, 1.15385\}\}$$

$$\{0.25+0.44441i, 0.25-0.44441i\}$$

For transformation of matrices, functions also are used:

Transpose[ ]—transpose of matrix
Det[ ]—calculation of matrix determinant
Tr[ ]—calculation of trace of matrix
Eigenvectors[m1]—calculation of matrix eigenvalues

The set of linear algebraic equations, represented in the matrix form, can be solved with the help of the **LinearSolve[ ]** function. Let us find the solution to the set of equations

$$0.3x_1 - 2.0x_2 = 5.0,$$

$$0.1x_1 + 0.2x_2 = -1.3.$$

We use this symbol to input the matrix:

$$\begin{pmatrix} \square & \square \\ \square & \square \end{pmatrix}$$

which is located on the toolbar. To input matrices and vectors of different sizes, it is necessary to choose the Mathematica menu: Input->Create Table/Matrix/Palette. and then determine the Number of rows and Number of columns. Solving the system of equations with matrix and vector,

$$m1 = \begin{pmatrix} 0.3 & -2.0 \\ 1.1 & 0.2 \end{pmatrix}$$

$$b1 = \begin{pmatrix} 5.0 \\ -1.3 \end{pmatrix};$$

with the help of the function

LinearSolve[m1,b1]

we obtain

$$\{\{-0.707965\},\{-2.60619\}\}$$

The solution to this set of equations could also be found using the inverse matrix

Inverse[m1].b1

The result will be the same.

It is necessary to note that, for addition and subtraction of matrices, the usual symbols are used. To multiply matrix by matrix, matrix by vector, and vector by vector (inner product of vectors), the dot symbol is used. To find the product of vector-column by vector-row, it is necessary to use the **Outer[ ]** function. Consider an example. Let us find the product of two vectors

cc={c1,c2};

dd={d1,d2};

Applying the function

Outer[Times,cc,dd]

yields

$$\{\{c1\,d1,c1\,d2\},\{c2\,d1,c2\,d2\}\}$$

The **MatrixExp[ ]** function is used in Mathematica for the calculation of matrix exponential. Let us consider the application of this function for solving the set of linear differential equations

$$\frac{dX}{dt} = AX$$

at the initial condition $X(0) = X_0$. The solution to such an equation has the form

$$X(t) = e^{At}X_0 \tag{1.1}$$

For matrix

$$A1 = \begin{pmatrix} 0.3 & -2.0 \\ 1.1 & 0.2 \end{pmatrix};$$

at the initial condition

$$x0 = \begin{pmatrix} -1.0 \\ 1.0 \end{pmatrix}$$

the solution to Equation (1.1) is obtained in the following way:

s1=Simplify[ComplexExpand[MatrixExp[A1*t].x0]]

$$\{\{e^{0.25t}((-1.+0.i)Cos[1.4824t]-(1.3829+0.i)Sin[1.4824t])\},$$

$$\{e^{0.25t}((1.+0.i)Cos[1.4824t]-(0.775771+0.i)Sin[1.4824t])\}\}$$

The **ComplexExpand[ ]** function, which expands expressions with complex numbers, is used for a solution's transformation. Items $0.i$ exist in the obtained solution. The function **Chop[ ]**, which in the general case allows the approximation of the real part of the number with the required precision, is used for the elimination of such items. Calculating

s2=Chop[s1]

yields

$$\{\{e^{0.25t}(-1.Cos[1.4824t]-1.3829Sin[1.4824t])\},$$

$$\{e^{0.25t}(1.Cos[1.4824t]-0.775771Sin[1.4824t])\}\}$$

For solving the nonhomogeneous matrix differential equation

$$\frac{dX}{dt} = AX + B \qquad (1.2)$$

we use the expression

$$X(t) = e^{At}X_0 + \int_0^t e^{A(t-\tau)}B(\tau)d\tau. \qquad (1.3)$$

When $B(\tau) = B = const$, then this expression can be represented as

$$X(t) = e^{At}X_0 + A^{-1}(e^{At} - I)B,$$

where A^{-1} is the inverse matrix; I is the unit matrix.

Let us find the solution to the Equation (1.2) for

$$A1 = \begin{pmatrix} -0.4 & -0.3 \\ 0.8 & -7.6 \end{pmatrix};$$

$$B1 = \begin{pmatrix} 10.0 \\ 0 \end{pmatrix};$$

$$X0 = \begin{pmatrix} 0 \\ 0 \end{pmatrix};$$

I2=IdentityMatrix[2];

At:=MatrixExp[A1*t];

X1=Simplify[At.X0+Inverse[A1].(At-I2).B1]

Then

$$\{\{23.1707 + 0.00620489e^{-7.56651t} - 23.1769e^{-0.433489t}\},$$

$$\{2.43902 + 0.148225e^{-7.56651t} - 2.58725e^{-0.433489t}\}\}$$

In these calculations the unit matrix of second order is determined with the help of the **Identity[2]** function. The function **At:=MatrixExp[A1*t]** is introduced for shortening the expressions.

1.4 Graphics Plotting

In Mathematica the application of various functions that enable the generation of 2D and 3D graphs, organized in various ways, is specified. The **Plot[ ]** function is used for plotting 2D graphs. Let us plot graphs of $y1 = aSin(\omega t)$ and $y2 = bt$ on the interval $t = 0.1 - 0.5$. Then, as a result,

ω=16.1;

y1=12.1*Sin[ω*t];

y2=8.7*t;

Plot[{y1,y2},{t,0.1,0.5},AxesLabel->{"t","y"}]

we obtain the graphs presented in Figure 1.1. The **Plot[ ]** function draws the graphs of functions presented in the list **{y1,y2}** at the interval **{t,0.1,0.5}**. In

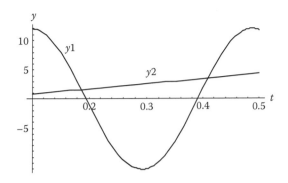

FIGURE 1.1
Graphs of $y1 = aSin(\omega t)$ and $y2 = bt$.

this example, the option used is **AxesLabel -> {"t", "y"}**, which establishes the labels to be put on the axes. Numerical values for ordinate axes are chosen by Mathematica after the calculation of all function values.

During the solving of differential equations, the obtained expressions are often presented as plots. Let us consider an example. We plot $x(t)$ and $y(t)$ functions, arising from the solution to the following set of differential equations:

$$eq1=-y'[t]-3*y[t]+x[t]+10;$$

$$eq2=2*x'[t]-1.8*x[t]-y[t];$$

$$s1=Simplify[DSolve[\{eq1==0,eq2==0,y[0]==-1,x[0]==2\},\{y[t],x[t]\},t]]//N;$$

$$Plot[y[t]/.s1,\{t,0,1.5\},AxesLabel->\{"t","y"\}]$$

$$Plot[x[t]/.s1,\{t,0,1.5\},AxesLabel->\{"t","x"\}]$$

Graphs of **y[t], x[t]** are presented in Figures 1.2 and 1.3.

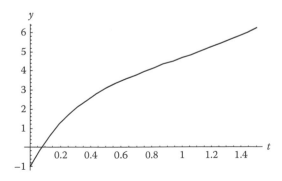

FIGURE 1.2
Graphs of function $y[t]$.

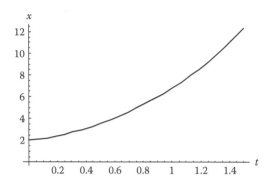

FIGURE 1.3
Graph of function $x[t]$.

The **ParametricPlot[ ]** function for making graphics of parametrically specified functions is used in Mathematica. Let us plot the graph of the functions specified parametrically with the help of $y1 = a_1 e^{-bt} Sin(\omega t)$ and $y2 = a_2 e^{-bt} Cos(\omega t)$. Then,

$$\omega = 60;$$

$$y1 = 12.1*Exp[-33*t]*Sin[\omega*t];$$

$$y2 = 2.4*Exp[-33*t]*Cos[\omega*t];$$

ParametricPlot[{y1,y2},{t,0.1,0.5},AxesLabel->{"y2","y1"},PlotRange->All]

The graph is shown in Figure 1.4.

When data are specified as a list, then it is necessary to use the **ListPlot[ ]** function for graphic presentation. Data can be represented either in the form

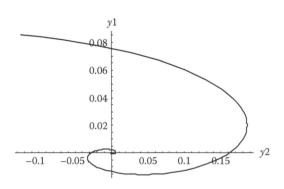

FIGURE 1.4
Graph of the functions specified parametrically.

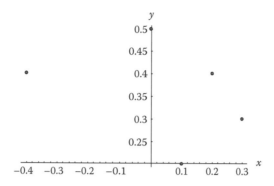

FIGURE 1.5
Graph of $y = f(x)$ in the form of points.

of $\{y1, y2,...\}$, or $\{\{x1, y1\}, \{x2, y2\}..\}$. In first case, for $y1$ $x1 = 1$, $y2$ $x2 = 2$, etc. In the second case, pairs of numbers correspond to values of points. For example, for the function $y = f(x)$, represented by the list

$$d1=\{\{0.1,0.2\},\{0.3,0.3\},\{0.2,0.4\},\{0.0,0.5\},\{-0.4,0.4\}\};$$

plotting of graphs is realized in the following way:

$$d1=\{\{0.1,0.2\},\{0.3,0.3\},\{0.2,0.4\},\{0.0,0.5\},\{-0.4,0.4\}\};$$

$$p1=ListPlot[d1,AxesLabel->\{"x","y"\},PlotStyle->\{PointSize[0.02]\}]$$

$$p2=ListPlot[d1,AxesLabel->\{"x","y"\},PlotJoined->True]$$

$$Show[p1,p2]$$

In Figure 1.5, the graph of the function in the form of points is presented. The Point size is established by the option **PlotStyle->{PointSize[0.02]}**. The minimum point size for a 2D graph is established Mathematica and is equal to 0.08.

Points can be joined by straight lines with the help of the **PlotJoined->True** option. This option is used for plotting the graph (Figure 1.6). The **Show[ ]** function draws two graphs together (Figure 1.7).

For making 3D plots in Mathematica the **Plot3D[ ]**, the **ParamericPlot3D[ ]** and **ListPlot3D[ ]** functions are used. For an application of the **Plot3D[ ]** function, let us consider an example. Let the functions have the form

$$z1=x+0.8*y;$$

$$z2=1.5*Sin[1.2*x]+2.0;$$

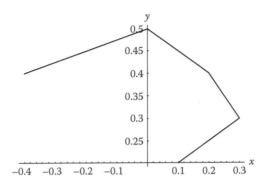

FIGURE 1.6
Graph of $y = f(x)$ in the form of straight-line segments.

Using the functions **Plot3D[ ]** and **Show[ ]**,

p1=Plot3D[z1,{x,0,4},{y,0,3},AxesLabel->{"x","y","z"},Shading->False];

p2=Plot3D[z2,{x,0,4},{y,0,3},Lighting->False];

Show[p1,p2]

we obtain graphs, which are shown in Figures 1.8, 1.9, and 1.10.
During plotting of the $z1 = f(x,y)$ function, we use the option **Shading->False,** which makes the surface white. The option **Lighting->False** allows drawing without an illumination.

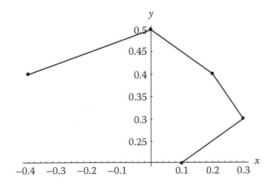

FIGURE 1.7
Graphs 1.5 and 1.6.

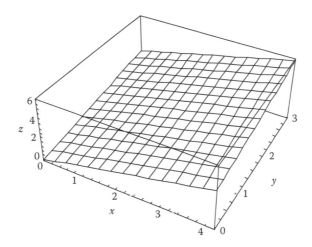

FIGURE 1.8
Graph of $z1 = f(x, y)$.

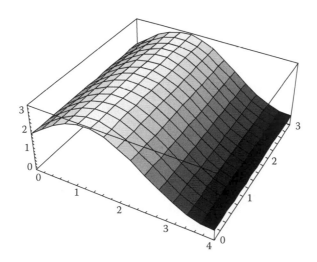

FIGURE 1.9
Graph of $z2 = \varphi(x, y)$.

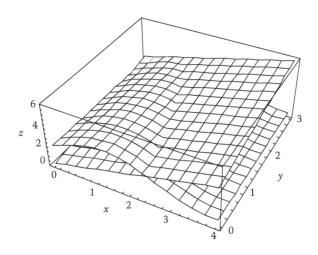

FIGURE 1.10
Graphs of $z1 = f(x,y)$ and $z2 = \varphi(x, y)$.

1.5 Overview of Elements and Methods of Higher Mathematics

In Mathematica there are derivate and integral operations. To calculate derivates **D[ ]** and **Dt[ ]**, functions are used. The function

$$D[a*Sin[b*x],x]$$

allows us to find the partial derivative $\frac{\partial}{\partial x}$:

$$abCos[bx]$$

The function

$$D[a*Sin[b*x],\{x,2\}]$$

allows us to find the second partial derivative:

$$-ab^2 Sin[b\,x]$$

The function

$$D[y*Sin[b*x]+y,x,y]$$

allows us to find the derivative $\frac{\partial}{\partial x} \frac{\partial}{\partial y}$:

$$bCos[bx]$$

In Mathematica, provision is made to define certain functions. For example,

f[x_]:=2.0*Exp[-x];

In the expression **f[x_]**, the argument **x_** points to the variable place, not to the variable itself. Using such a function's determination, the derivative calculation

D[f[t],t]

gives the following expression:

$$-2 \cdot e^{-t}$$

To calculate the total derivatives and the differential, the **Dt[ ]** function is used. For example, as a result of the calculation

Dt[a1*x]

we obtain

$$xDt[a1]+a1Dt[x]$$

There are analytic and numerical methods for calculating integrals in Mathematica. For the indefinite integral, calculation is made by the function defined by the symbol

$$\int \Box d\Box$$

for example,

$$\int Cos[b*x]dx$$

or the function defined by the name **Integrate[ ]**, for example,

Integrate[Cos[b*x],x]

As a result of indefinite integral calculation, we obtain

$$\frac{Sin[b\,x]}{b}$$

For definite integral calculation, there are also two applicable forms. For example, calculating the integral with the help of one of the forms

$$\int_0^1 Exp[-b*x]\,dx$$

Integrate[Exp[−b*x], {x,0,1}]

we obtain the same result:

$$\frac{1}{b} - \frac{e^{-b}}{b}$$

For numerical integration of the expressions, the **NIntegrate[ ]** function is used. Consider the following example. Find the integral of a function

$$\frac{1}{b+x+\sin x}$$

Calculating indefinite integral

f[x_]:=1/(b+x+Sin[x]);

Integrate[f[x],x]

we obtain

$$\int \frac{1}{b+x+Sin[x]}\,dx$$

Mathematica shows that this indefinite integral cannot be calculated. The numerical value of this integral for $b = 2.2$ and the interval 0–1 is calculated in the following way:

B=2.2;

NIntegrate[f[x],{x,0,1}]

Then,

0.326247

In solving various problems, functions very often are presented as a sum. For the Taylor series expansion, the **Series[ ]** function is used. For example, the Taylor series of the function

$$\frac{1}{2+t}$$

up to 3-d order is found in the following way:

$$s1=Series[1/(2+t),\{t,0,3\}]$$

Then,

$$\frac{1}{2}-\frac{t}{4}+\frac{t^2}{8}-\frac{t^3}{16}+O[t]^4$$

For series truncating, the **Normal[ ]** function is used. Using this function

$$s2=Normal[s1]$$

we obtain

$$\frac{1}{2}-\frac{t}{4}+\frac{t^2}{8}-\frac{t^3}{16}$$

In Mathematica there are functions that are used for finding the Fourier transform, Laplace transform, and Z-transform. The Fourier transform is determined by the function **FourierTransform[ ]**. For example, for function

$$f(t)=\begin{cases} e^{-t}, & t>0, \\ 0, & t\leq 0 \end{cases}$$

the Fourier transform

$$f1[t\_]:=Exp[-t]*UnitStep[t];$$

$$FourierTransform[f1[t],t,\omega]$$

gives the expression

$$\frac{i}{\sqrt{2\pi}\,(i+\omega)}$$

The **f1[t_]** function is defined by the unit step function **UnitStep[t]**. The inverse Fourier transform of the function

$$\frac{1}{3+i\omega}$$

is determined with the help of the function

$$InverseFourierTransform[1/(3+I*\omega),\omega,t]$$

Then,

$$e^{3t} \sqrt{2\pi} UnitStep[-t].$$

The Laplace transform and Z-transform are applied similarly.

1.6 Use of the Programming Elements in Mathematical Problems

In Mathematica the use of defined if-statements and functions allow effective organization of the process of calculation of complex expressions. An if-statement has the form **If[ ]**. Let us consider an example in which it is necessary to calculate the integral of a function

$$f(t) = \begin{cases} e^{-t}, & t > 0, \\ t+1, & t \leq 0 \end{cases}$$

Using an if-statement, determine the function in the following way:

f[t_]:=If[t>0,Exp[-t],t+1];

The graph of this function

Plot[f[t],{t,-2,2},AxesLabel->{"t","f"}]

is presented in Figure 1.11

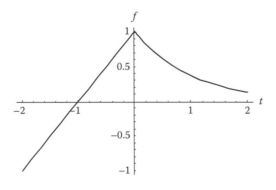

FIGURE 1.11
Graph of *f(t)*.

The integral of function

$$\text{Integrate[f[x],\{x,-1,1\}]}$$

is equal to

$$\frac{3}{2} - \frac{1}{e}$$

For a finite series sum calculation, it is expedient to use the **For[ ]** function, by the help of which loops are created in the program. For example, the sum of numbers $2n$ for $n = 1\ldots100$ can be found as follows:

$$\text{i1=0;}$$

$$\text{For[n=1,n≤100,i1=i1+2*n;n++];}$$

$$\text{i1}$$

Then,

$$10110$$

In this expression, **n=1** corresponds to the initial value, but **n≤100**, corresponds to the finite value of the variable. The expression **n++** shows that the variable increases by 1.

In another example we consider the finite series formation for the function $\frac{1}{1+an}$. As a result of using the **For[ ]** function

$$\text{i1=0;}$$

$$\text{For[n=1,n≤4,i1=i1+1/(1+a*n);n++];}$$

$$\text{i1}$$

we obtain

$$\frac{1}{1+a} + \frac{1}{1+2a} + \frac{1}{1+3a} + \frac{1}{1+4a}$$

We may obtain the same result using the **Sum[ ]** function. To form the finite series, we should write

$$\text{Sum[1/(1+a*n),\{n,1,4\}]}$$

It is expedient to use the **For[ ]** function for repeating operations with matrices and vectors. For example, let us find the product

$$A^3B = (A(A(Ab))),$$

where

$$A = \begin{pmatrix} -1.2 & -0.7 \\ 2.0 & -0.9 \end{pmatrix}; \quad B = \begin{pmatrix} 4.1 \\ -6 \end{pmatrix}.$$

The calculation of the product is made as follows:

$$\mathbf{A1} = \begin{pmatrix} -1.2 & -0.7 \\ 2.0 & -0.9 \end{pmatrix};$$

$$\mathbf{B1} = \begin{pmatrix} 4.1 \\ -6 \end{pmatrix};$$

C1=B1;

For[n=1,n<=3,C1=A1.C1;n++];

C1

Then,

$$\{\{19.9632\}, \{-5\}\}$$

The same results can be obtained with the help of the **Do[ ]** function. With the presence of a condition, repeating calculations can be realized by means of the **While[ ]** function.

2

Calculation of Transition and Steady-State Processes

2.1 Calculation of Processes in Linear Systems

Electromagnetic processes in linear systems are described by linear differential equations with constant coefficients:

$$\frac{dX}{dt} = AX + Be(t),\qquad(2.1)$$

where X is the vector of state variables, $e(t)$ is the forcing function, A is the matrix, and B is the vector with constant elements.

Let us present the solution to Equation 2.1 in the form

$$X(t) = e^{At}X_0 + \int_0^t e^{A(t-\tau)}Be(\tau)\,d\tau,\qquad(2.2)$$

where X_0 is the initial condition of the vector X at $t=0$.

As an example, we consider the circuit represented in Figure 2.1.

Using Kirchhoff's laws we may write the differential equations for current i and voltage u in the following way:

$$e(t) = iR_1 + L\frac{di}{dt} + u;$$

$$i = C\frac{du}{dt} + \frac{u}{R_2}.$$

Transforming these equations to the normal form (Equation 2.1), matrix A and vector B are

$$A = \begin{vmatrix} -\dfrac{R_1}{L} & -\dfrac{1}{L} \\[2mm] \dfrac{1}{C} & -\dfrac{1}{R_2 C} \end{vmatrix}; \quad B = \begin{vmatrix} \dfrac{1}{L} \\[2mm] 0 \end{vmatrix}.$$

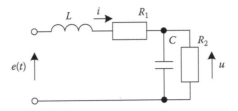

FIGURE 2.1
Circuit with linear elements.

2.1.1 Solution by the Analytical Method

In Mathematica the process of solution begins with data input:

$$R1=0.2;$$

$$R2=8.0;$$

$$L1=0.008;$$

$$C1=8.0*10^{\wedge}(-4);$$

$$A1 = \begin{pmatrix} -R1/L1 & -1/L1 \\ 1/C1 & -1/(R2*C1) \end{pmatrix};$$

$$B1 = \begin{pmatrix} 1/L1 \\ 0 \end{pmatrix};$$

$$f=50.0;$$

$$\omega=2*\pi*f;$$

$$e[\tau\_]:=20.0*Sin[\omega*\tau];$$

$$X0 = \begin{pmatrix} 0 \\ 0 \end{pmatrix};$$

Remember that matrix and vector of arbitrary dimensions are inputted by Input->Create Table/Matrix/Palette… Then, in the opened window, the number of rows (Number of rows) and columns (Number of columns) are set.

In a row

$$e[\tau\_]:=20.0*Sin[\omega*\tau];$$

the user-defined function is determined for a variable τ. Sign := shows that the right part of the expression is not calculated and is not generated in the output row.

In the next cell the function **At1[t_]** is defined, and the solution **XT** is determined as follows:

$$\textbf{At1[t\_]:=MatrixExp[A1*t];}$$

$$\textbf{XT=Chop[ComplexExpand[At1[t].X0+At1[t].}$$

$$\textbf{Integrate[(At1[-\tau].B1)*e[\tau],\{\tau,0,t\}]]];}$$

In the first row the expression **MatrixExp[A1*t]** defines a function for the matrix exponent of matrix **A1**. In the next row, the solution (Equation 2.2) is determined. In this expression the "dot" symbol points to the matrix multiplication or matrix by vector multiplication. The function **Integrate[(At1[-τ].B1)*e(τ),{τ,0,t}]** finds the defined integral of the function **(At1[-τ].B1)*e(τ)[τ]** with respect to the variable **τ** determined on the interval **0-t**.

The graphs are plotted with the help of the function

$$\textbf{Plot[\{XT[[2]],e[t],XT[[1]]\},\{t,0,0.02\},AxesLabel\_\{"t","u i"\}]}$$

The **Plot[]** function plots the graphs of the functions, which are represented in the list **{XT[[2]],e[t],XT[[1]]}**. Time diagrams are shown in Figure 2.2. The argument **t** changes from 0 to 0.02. The argument and its change are written as a list **{t,0,0.02}**. During **XT** calculation, Mathematica determines itself that this expression is a vector and calculates its dimension. The extraction of the vector element is produced by means of writing **XT[[1]]**, that is, the first element of the vector, which determines the current in this case, is chosen. The option **AxesLabel->{"t","u i"}** points to the necessity of output of symbols **t** and **u i** along the abscissa and ordinate axes. Numeral values for the ordinate axis are chosen by Mathematica® after the calculation of all function values.

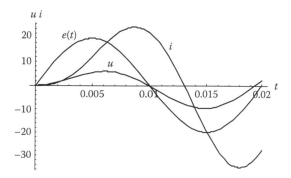

FIGURE 2.2
Input voltage *e(t)*, inductor current *i*, and capacitor voltage *u* responses (*e(t)* and *u* in volts, *i* in amperes, time *t* in seconds).

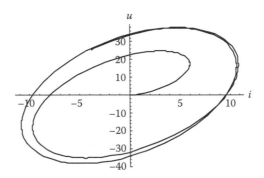

FIGURE 2.3
The phase-plane trajectory (u in volts, i in amperes).

In a row,

**ParametricPlot[{Part[XT[[1]],1],Part[XT[[2]],1]},{t,0,0.05},PlotRange->All,
AxesLabel->{"i","u"},DefaultFont->{"Arial",12}]**

the **ParametricPlot** function determines the parametric graph (Figure 2.3), which corresponds to the phase-plane portrait. The graph is given by the arguments {**Part[XT[[1]],1],Part[[2]],1**} and is drawn for the interval {**t,0,0.05**}. The option **PlotRange->All** points to the necessity for the output of all calculated points in the picture.

The calculation process of the whole notebook is produced by the choice Kerner->Evaluation->Evaluate Notebook. If it is necessary to calculate a cell in which the cursor is situated, one needs to press keys Shift and Enter at the same time. Remember that pressing only the Enter key leads to a line feed.

The other way of finding a solution for differential equations is based on the use of the **DSolve[ ]** function

**sol1=Chop[ComplexExpand[DSolve[{i'[t]==-R1/L1*i[t]-1/L1*u[t]+e[t]/
L1,u'[t] == 1/C1*i[t]-1/(R2*C1)*u[t]},i[0]==0,u[0]==∷0},{i[t],u[t]},t]]]**

In this case, Mathematica tries to find the analytical solution to the set of differential equations. Since the symbol ';' is absent, the expression is generated in the output cell (as a list):

$$\{\{i[t] \to (4.37655+1.85393i)e^{(-90.625-779.598i)t}((-0.718315-0.695718i)e^{389.799it}$$

$$+1.ie^{1169.4it}+(0.718315-0.304282i)e^{(90.625+779.598i)t}Cos[314.159t]$$

$$+(1.89678-0.803488i)e^{(90.625+779.598i)t}Sin[314.159t]),$$

$$u[t] \to (12.6745+8.07898i)e^{(-90.625-779.598i)t}((0.422167-0.906518i)e^{389.799t}+1.e^{1169.4it}$$

$$-(1.42217-0.906518i)e^{(90.625+779.598i)t}Cos[314.159t]$$

$$+(1.53503-0.978459i)e^{(90.625+779.598i)t}Sin[314.159t])\}\}$$

In the list {i[t],u[t]} of the **DSolve[ ]** function, related variables are defined by which the solution is found and, at the end of this function, also the independent variable **t** is defined. The graph is plotted with the help of the function

ParametricPlot[{Part[i[t]/.sol1,1],Part[u[t]/.sol1,1]},{t,0,0.05},

PlotRange->All,AxesLabel->{"i","u"},DefaultFont->{"Arial",12}];

As a result we obtain the plane-phase portrait analogous to the one shown in Figure 2.3.

2.1.2 Solution by the Numerical Method

Let us use the numerical method of Mathematica for the solution of the system (Equation 2.1). In the row

sol2=NDSolve[{i′[t]==-R1/L1*i[t]-1/L1*u[t]+e[t]/L1,u′[t]==1/C1*i[t]

-1/(R2*C1)*u[t],i[0]==0,u[0]==0},{i[t],u[t]},{t,0,0.05}];

the numerical solution is given for the **sol2** variable and, in the next output cell, an interpolation polynomial is defined:

{{i[t] -> InterpolatingFunction[{{0., 0.05}}, "<>"][t],

u[t] -> InterpolatingFunction[{{0., 0.05}}, "<>"][t]}}

The equations set and initial conditions of variables are specified in the form of the list for the **NDSolve[ ]** function

{i′[t]==-R1/L1*i[t]-1/L1*u[t]+e[t]/L1,u′[t]==1/C1*i[t]-1/(R2*C1)*

u[t],i[0]==0,u[0]==0},

Further, the variables are defined in the form of a list and, at the end, the list of the independent variable **t** and its range **{t,0,0.05}** are specified.

For plotting of the graph we use the function

**ParametricPlot[{Part[Evaluate[i[t]/.sol2],1],Part[Evaluate[u[t]/.
sol2],1]},{t,0,0.05},**

PlotRange->All,AxesLabel->{"i","u"},DefaultFont->{"Arial",12}];

The expression **[i[t]/.sol2]** shows that the value of the **sol2** solution must be substituted for the current **i[t]**. The **Evaluate[ ]** function shows that the expression must be calculated. The **Part[,1]** function chooses the first expression from the list, that is, allows the cancellation of braces. The

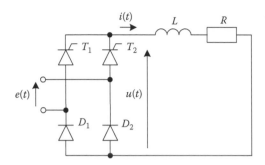

FIGURE 2.4
Topology of the thyristor-controlled rectifier.

ParametricPlot function outputs the graph of the phase-plane portrait similarly to that in Figure 2.3.

2.2 Calculation of Processes in the Thyristor Rectifier Circuit

Let us determine a steady-state process in the circuit of the semicontrolled rectifier (Figure 2.4). Thyristors are turned on by periodical impulses, but impulses for thyristor T_1 are shifted by half of the period from impulses for thyristor T_2.

We assume that the current through the inductor is continuous, the inductor is a linear element, and that an ideal switch model for diodes and thyristors is used. The example of the time diagram of the voltages is shown in the Figure 2.5.

Processes in this rectifier can be described by the differential equation

$$L\frac{di(t)}{dt} + Ri(t) = u(t), \tag{2.3}$$

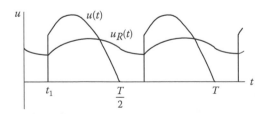

FIGURE 2.5
Processes in the thyristor rectifier circuits.

where

$$
u(t) = \begin{cases} 0, & nT/2 \le t \le t_1 + nT/2; \\ E\,|\sin(\omega t)|, & t_1 + nT/2 \le t \le (n+1)T/2; \end{cases}
$$

$n = 0,1,2,..;$ $\omega = \frac{2\pi}{T}$, T is the period of the supply voltage $e(t)$; and t_1 is the turn-on time of the thyristors.

In order to show features of a method in more detail, we shift the ordinate axis at the point $t = t_1$. Then the voltage $u(t)$ takes the form

$$
u(t) = \begin{cases} E\,|\sin(\omega t + t_1)|, & nT/2 \le t \le nT/2 - t_1; \\ 0, & nT/2 - t_1 \le t \le (n+1)T/2; \end{cases}
\tag{2.4}
$$

Since processes in such a circuit are described by a stationary differential equation, we can use the Laplace transform. Applying the Laplace transform to Equation 2.3 with the voltage (Equation 2.4), one obtains the following equation:

$$
(pL + R)I(p) = U(p),
\tag{2.5}
$$

where $I(p)$ is the Laplace transform of the current $i(t)$; and $U(p)$ is the Laplace transform of the voltage $u(t)$. At the same time we assume that the initial condition of the current $i(t)$ is equal to zero. The right part of this equation is obtained by taking into account that the voltage $u(t)$ is periodic, with the period equaling $T/2$. The transform of a periodic function $f(t) = f(t+T)$ is given by

$$
F(p) = \frac{\int_0^T f(t)e^{-pT}\,dt}{1 - e^{-pT}}.
$$

Let us use Mathematica for deriving the expression for the transform $U(s)$. In the cell we evaluate the nominator of the function $F(p)$:

$$
\mathbf{Ee1 = FullSimplify}\left[\left. \int_0^{T/2 - t1} \mathbf{E1 * Sin[}\ * (t + t1)] * \mathbf{Exp[-p * t]}\,\right/ .\mathbf{T} - > 2 * \mathbf{Pi/}\right]
$$

Mathematica outputs the expression

$$
E1\left(\frac{e^{p\left(t1 - \frac{\pi}{\omega}\right)}\omega + \omega\,\mathrm{Cos}[t1 * \omega] + p\,\mathrm{Sin}[t1 * \omega]}{p^2 + \omega^2}\right)
$$

Solving Equation 2.5 for $I(p)$ yields

$$I(p) = \frac{E\left[e^{p\left(t_1 - \frac{\pi}{\omega}\right)}\omega + \omega\cos[t_1\omega] + p\sin[t_1\omega]\right]}{(pL+R)(p^2+\omega^2)\left(1 - e^{-p\frac{T}{2}}\right)}. \tag{2.6}$$

Natural and forced responses of the current could be determined by using the inverse Laplace transform, and could be expressed thus:

$$i(t) = i_n(t) + i_f(t), \tag{2.7}$$

where $i_n(t)$ is the natural response; and $i_f(t)$ is the forced response. A forced response is also called a steady-state process. These responses are determined by calculating residues with respect to all poles of the transform $I(s)$ as follows:

$$i_n(t) = \sum_{k=1}^{K} \operatorname{Re} s[I(p)e^{pt}, p_k];$$

$$i_f(t) = \sum_{l} \operatorname{Re} s[I(p)e^{pt}, p_l],$$

where p_k are the poles of a transfer function $\frac{1}{pL+R}$; K is the order of the differential equation describing the circuit; and p_l are the poles of the forced function, that is, poles of the function $\frac{1}{(p^2+\omega^2)(1-e^{-p\frac{T}{2}})}$. Since the function $1 - e^{-p\frac{T}{2}}$ has infinitely many roots

$$p_m = \pm j\frac{4\pi m}{T}, \quad m = 0,1,2,\ldots$$

then the steady-state solution has infinitely many terms.

Let us consider a method (Waidelich, 1946) that allows finding a steady-state process without using a periodicity condition. The method is based on introducing a continuous function $u_c(t)$ (Rudenko et al., 1980) that coincides with the forced function $u(t)$ on the interval where the steady-state process is determined (Figure 2.6).

We consider the equation

$$L\frac{di_c(t)}{dt} + Ri_c(t) = u_c(t)$$

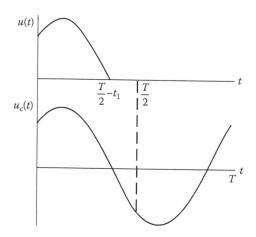

FIGURE 2.6
Forced and continuous functions.

which differs from Equation 2.3 only by the right-hand part. The Laplace transform of this equation gives

$$(pL+R)I_c(p)=U_c(p),$$

where $U_c(p)$ is the Laplace transform of the $u_c(t)=E\sin(\omega(t+t_1))$ voltage; and $I_c(p)$ is the Laplace transform of a current corresponding to the voltage $U_c(p)$; $U_c(p)=\frac{E(p\sin\omega t_1+\omega\cos\omega t_1)}{p^2+\omega^2}$. Solving this equation for $I_c(p)$ yields

$$I_c(p)=\frac{E(p\sin\omega t_1+\omega\cos\omega t_1)}{(pL+R)(p^2+\omega^2)},\qquad(2.8)$$

Using the inverse Laplace transform, we obtain from Equation 2.8 the solution

$$i_c(t)=\tilde{i}_n(t)+\tilde{i}_f(t),\qquad(2.9)$$

where $\tilde{i}_n(t)$ is the natural response; and $\tilde{i}_f(t)$ is the forced response. These responses can be determined by calculating residues with respect to all poles of the transform $I_c(p)$:

$$\tilde{i}_n(t)=\sum_{k=1}^{K}\operatorname{Res}\left[I_c(p)e^{pt},p_k\right];$$

$$\tilde{i}_f(t)=\sum_{q=1}^{Q}\operatorname{Res}\left[I_c(p)e^{pt},p_q\right],$$

where p_k are the poles of the transfer function $\frac{1}{pL+R}$; p_q are the poles of the forced function $U_c(p)$; and Q is the number of poles of the forced function $U_c(p)$.

Since forced functions $u(t)$ and $u_c(t)$ equal on the interval $0-(\frac{T}{2}-t_1)$, solutions to (2.7) and (2.9) equal each other on the same interval:

$$i_n(t)+i_f(t)=\tilde{i}_n(t)+\tilde{i}_f(t),$$

Therefore, one can write

$$i_f(t)=\tilde{i}_n(t)+\tilde{i}_f(t)-i_n(t), \tag{2.10}$$

In this expression the steady-state process is described by a sum of finite terms.

Let us use Mathematica for deriving a solution. In a cell we introduce expressions (2.6) and (2.8)

Iu:=Ee1/(p*L+R)/(1-Exp[-p*T/2]);

Ic:=E1*(p*Sin[ω*t1]+ω*Cos[ω*t1])/(p^2+ω^2)/(p*L+R);

In this cell, **Iu** corresponds to $I(p)$, and **Ic** corresponds to $I_c(p)$. In the next cell we find the inverse Laplace transform by evaluating the residues:

α=R/L;

p1=Iω;

in1=Residue[Iu*Exp[p*t],{p, α}]

icf1=Simplify[Factor[ExpToTrig[Residue[Ic*Exp[p*t],{p,p1}]+

Residue[Ic*Exp[p*t],{p,-p1}]]]]

Icn1=Residue[Ic*Exp[p*t],{p,- α}]

In this cell, **in1** corresponds to $i_n(t)$, **icf1** corresponds to $\tilde{i}_f(t)$, **icn1** corresponds to $\tilde{i}_n(t)$, α is the pole of the transfer function $\frac{1}{pL+R}$, and **p1** is the pole of the function $\frac{1}{p^2+\omega^2}$. Mathematica outputs the following expressions:

$$\frac{e^{-\frac{Rt}{L}-\frac{R\left(t1-\frac{\pi}{\omega}\right)}{L}}E1\left(-L\omega+e^{\frac{R\left(t1-\frac{\pi}{\omega}\right)}{L}}(-L\omega Cos[t1\omega]+RSin[t1\omega])\right)}{\left(-1+e^{\frac{RT}{2L}}\right)(R^2+L^2\omega^2)}$$

$$\frac{E1(-L\omega Cos[(t+t1)\omega]+RSin[(t+t1)\omega])}{R^2+L^2\omega^2}$$

$$-\frac{e^{\frac{Rt}{L}}E1(-L\omega Cos[t1\omega]+RSin[t1\omega])}{R^2+L^2\omega^2}$$

Simplifying the right part of (2.10), one obtains

$$i_f(t) = \frac{E}{\sqrt{R^2 + \omega^2 L^2}} \left[\frac{\left(e^{-\alpha t_1} \sin\varphi + \sin(\varphi - \omega t_1)\right)}{\left(e^{\alpha \frac{T}{2}} - 1\right)} e^{-\alpha\left(t - \frac{T}{2}\right)} + \sin(\omega t - \varphi) \right],$$

where

$$\alpha = \frac{R}{L}; \quad \varphi = acrtg\left(\frac{\omega L}{R}\right).$$

Now we determine a solution on the second interval. In order to simplify the calculation, it would be expedient to shift the ordinate axis at the point $t = T/2 - t_1$. However, this is the same as that we find for the forced function in Figure 2.5. So, we find the solution on the interval $0 - t1$.

Using Mathematica we find a Laplace transform for the voltage $u(t)$ defined as in (2.3):

$$\text{Ee2} = \text{FullSimplify}\left[\int_{t1}^{T} E1 * \text{Sin}[\quad * t] * \text{Exp}[-p * t]\right]/.T -> 2 * \text{Pi}/$$

Mathematica outputs the expression

$$\frac{E1\left(e^{-\frac{\pi p}{\omega}}\omega + e^{-pt1}(\omega\,\text{Cos}[t1 * \omega] + p\,\text{Sin}[t1 * \omega])\right)}{p^2 + \omega^2}$$

In the interval $0 - t1$, the voltage $u(t)$ is equal to zero. Therefore, the continuous function $u_c(t)$ equals zero, and the Laplace transform has the same value, that is, $U_c(p) = 0$. In that case, a natural response $\tilde{i}_n(t)$ and a forced response $\tilde{i}_f(t)$ are equal to zero.

We input the expression of the solution to Equation 2.5 with **Ee2**:

Iu2:=Ee2/(p*L+R)/(1-Exp[-p*T/2]);

Then we determine a solution corresponding to the forced function by calculating residue

in2=Residue[Iu2*Exp[p*t],{p,-α}]/.t->(t-T/2+t1)

In this row we substitute t by $t - (T/2 - t_1)$, which allows us to carry the solution at the point $t = (T/2 - t_1)$.

Mathematica outputs the expression

$$\frac{e^{-\frac{Ru}{L}-\frac{R\left(t1-\frac{T}{2}\right)}{L}}E1\left(-e^{\frac{\pi R}{L\omega}}L\omega+e^{\frac{Rt1}{L}}(-L\omega Cos[t1\omega]+RSin[t1\omega])\right)}{\left(-1+e^{\frac{RT}{2L}}\right)(R^2+L^2\omega^2)}$$

Now we enter the values of parameters of the circuit:

$$E1=310.0;$$

$$R=20.0;$$

$$L=0.04;$$

$$t1=2*10^(-3);$$

$$T=20*10^(-3);$$

$$\omega=2*Pi/T;$$

The graphs of the current of the steady-state process form with the help of the functions

p1i=Plot[icf1+icn1-in1,{t,0,T/2-t1},AxesLabel->{"t","i"},PlotRange->{0,15},
DisplayFunction->Identity];

p2i=Plot[-in2,{t,T/2-t1,T/2},AxesLabel->{"t","i"},
DisplayFunction->Identity];

Since the solutions on the second interval for the continuous function are equal to zero, in the last row one uses $i_f(t)=-i_n(t)$.

The graphical output using the function

Show[{p1i,p2i},DisplayFunction->$DisplayFunction]

is presented in Figure 2.7. The option **DisplayFunction->Identity** forms a graphical object but suppresses output. All characteristics are plotted simultaneously with the help of the **Show[]** function. The option **DisplayFunction->$DisplayFunction** allows the display of the graphical object.

It should be noted that the considered method does not depend on an analyzed circuit. The circuit must be described by linear stationary differential equations.

For such a linear stationary system we can determine the average value and harmonics of the steady-state process of the current. Let us express a function $f(t)$ by the complex Fourier series

$$f(t)=\sum_{n=-\infty}^{\infty}c_ne^{jnt},$$

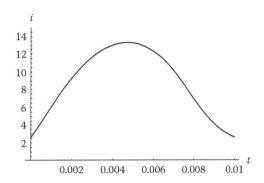

FIGURE 2.7
Steady-state process of the current (i in amperes, time t in seconds).

where c_n are the Fourier coefficients. One can see that, in this expression, the coefficients are multiplied by e^{jnt}. Comparing this expression with the inverse Laplace transform, we see that, in order to determine coefficients c_n, it is necessary to evaluate the residues of the $I(p)$ transform with respect to the poles of the $\frac{1}{(p^2+\omega^2)(1-e^{-p\frac{T}{2}})}$. These poles are $0, \pm j\omega, \pm j2\omega, \pm j4\omega, \ldots$

Let us use Mathematica for deriving the average value and harmonics. The average value is obtained by calculating the residue with respect to $s = 0$:

$$\textbf{Residue[Iu,\{p,0\}]}$$

Mathematica outputs the expression

$$\frac{E1\left(1+\text{Cos}\left[\frac{2\pi t1}{T}\right]\right)}{\pi R}$$

It should be noted that the foregoing and other functions of Mathematica that follow are to be used before inputting the parameter values.

For calculating the first harmonic $s = \pm j\omega$, we input the expression

$$\textbf{Residue[Iu,\{p,I*}\omega\textbf{\}]}$$

As a result, we obtain zero.

For calculating the second harmonic $p = \pm j2\omega$, we evaluate the residues as follows:

c2=FullSimplify[ExpToTrig[Residue[Iu,{p,I*4*Pi/T}]]/.ω->2*Pi/T]]

c2c=FullSimplify[ExpToTrig[Residue[Iu,{p,–I*4*Pi/T}]]/. ω->2*Pi/T]]

Mathematica outputs expressions

$$\frac{jE1T\left(\text{Cos}\left[\frac{2\pi t1}{T}\right]+\text{Cos}\left[\frac{4\pi t1}{T}\right]+8\,j\text{Cos}\left[\frac{\pi t1}{T}\right]^{3}\text{Sin}\left[\frac{\pi t1}{T}\right]\right)}{3\pi(4L\pi-jRT)}$$

$$-\frac{jE1T\left(\text{Cos}\left[\frac{2\pi t1}{T}\right]+\text{Cos}\left[\frac{4\pi t1}{T}\right]-8\,j\text{Cos}\left[\frac{\pi t1}{T}\right]^{3}\text{Sin}\left[\frac{\pi t1}{T}\right]\right)}{3\pi(4L\pi+jRT)}$$

The Fourier coefficients a_n, b_n, and an amplitude of the harmonic are calculated using

$$a_n = c_n + c_{-n};$$

$$b_n = j(c_n - c_{-n});$$

$$\sqrt{a_n^2 + b_n^2}$$

Computing the following expressions

an=Simplify[c2+c2c];

bn=Simplify[I*(c2-c2c)];

Simplify[Sqrt[an^2+bn^2]]

yields

$$4\sqrt{-\frac{\dfrac{E1^{2}T^{2}\text{Cos}\left[\frac{\pi t1}{T}\right]^{4}\left(-5+4\text{Cos}\left[\frac{2\pi t1}{T}\right]\right)}{16L^{2}\pi^{2}+R^{2}T^{2}}}{3\pi}}$$

In the same way we can calculate the other harmonics.

2.3 Calculation of Processes in Nonstationary Circuits

Let us consider the calculating procedure of processes in the open-loop system with the Boost converter as shown in Figure 2.8. The periodical pulses of an independent generator with a period T and duration t_1 are fed to the base of the transistor. On the interval $nT \le t \le nT + t_1$ ($n = 0, 1, 2, \ldots$), the transistor is opened. The current of the power supply flows through the inductor and transistor, and the capacitor is discharged through the resistor. On the

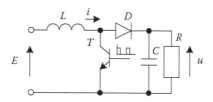

FIGURE 2.8
The topology of the Boost converter.

interval $nT+t_1 \le t \le (n+1)T$, the transistor is closed. The inductor maintains the current, which flows through the power supply, diode, and RC-circuit.

With enough precision for system modeling, the transistor and diode can be presented by the RS model, that is, as a switch with a resistance. We assume that the current through the inductor is continuous, and that the inductor and capacitor are linear elements. In the on state, the transistor and diode have equal resistances.

The equivalent circuit for the interval $nT \le t \le nT+t_1$ is presented in Figure 2.9. The electromagnetic processes are described by the matrix differential equation

$$\frac{dX(t)}{dt} = A_1 X(t) + B_1 E , \qquad (2.11)$$

where

$$X(t) = \begin{vmatrix} i \\ u \end{vmatrix} ; \quad A_1 = \begin{vmatrix} -\dfrac{R_1}{L} & 0 \\ 0 & -\dfrac{1}{RC} \end{vmatrix} ; \quad B_1 = \begin{vmatrix} \dfrac{1}{L} \\ 0 \end{vmatrix} ; \quad R_1 = R_i + R_t ;$$

R_t is the resistance of the transistor in on state, and R_i is the resistance of the inductor.

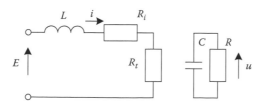

FIGURE 2.9
The equivalent circuit of the converter. The transistor is on, and the diode is off.

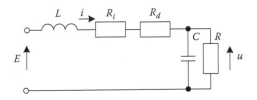

FIGURE 2.10
The equivalent circuit of the converter. The transistor is off, and the diode is on.

The equivalent circuit for the interval $nT + t_1 \leq t \leq (n+1)T$ is presented in Figure 2.10.

The electromagnetic processes are described by the matrix differential equation

$$\frac{dX(t)}{dt} = A_2 X(t) + B_2 E,$$ (2.12)

where

$$A_2 = \begin{vmatrix} -\dfrac{R_2}{L} & -\dfrac{1}{L} \\[2mm] \dfrac{1}{C} & -\dfrac{1}{RC} \end{vmatrix}; \quad B_2 = B_1; \quad R_2 = R_i + R_d;$$

R_d is the resistance of the diode in the on state; $R_2 = R_1$.

For the solution to Equations 2.11 and 2.12 we use the expression (2.2). Since $Be(t) = B_1 E$ does not depend on time, we can take the integral in (2.2) and write as follows:

$$X(t) = e^{A_1(t-nT)} X(nT) + A_1^{-1}(e^{A_1(t-nT)} - I)B_1 E,$$

$$X(t) = e^{A_2(t-nT-t_1)} X(nT + t_1) + A_2^{-1}(e^{A_2(t-nT-t_1)} - I)B_1 E,$$ (2.13)

where A_1^{-1}, A_2^{-1} are the inverse matrices; I is the unit matrix; $X(nT)$ is the initial condition of the vector $X(t)$ for the interval $nT \leq t \leq nT + t_1$; and $X(nT + t_1)$ is the initial condition of the vector $X(t)$ for the interval $nT + t_1 \leq t \leq (n+1)T$.

The solution for the system is based on the consequent use of expressions (2.13). In addition, the initial conditions $X(nT)$ and $X(nT + t_1)$ are determined from the vector $X(t)$ at the end of corresponding intervals:

$$X(nT + t_1) = e^{A_1 t_1} X(nT) + A_1^{-1}(e^{A_1 t_1} - I)B_1 E,$$

$$X((n+1)T) = e^{A_2(T-t_1)} X(nT + t_1) + A_2^{-1}(e^{A_2(T-t_1)} - I)B_1 E.$$ (2.14)

In the steady state, $X(nT) = X((n+1)T)$. Using this condition and the set (2.14), we obtain

$$X(nT) = (I - e^{A_2(T-t_1)}e^{A_1 t_1})^{-1}\left[e^{A_2(T-t_1)}A_1^{-1}(e^{A_1 t_1} - I) + A_2^{-1}(e^{A_2(T-t_1)} - I)\right]B_1 E. \quad (2.15)$$

Let us consider how to use Mathematica for determination of the transition and steady-state behaviors. In the first cell, the values of the parameters are inputted:

R1=4.0;

L1=0.02;

C1=1.0*10^(-5);

R2=15.0;

E1=20;

t1=0.000469;

T=1.0*10^(-3);

$$A1 = \begin{pmatrix} -R1/L1 & 0 \\ 1/C1 & -1/(R2 * C1) \end{pmatrix};$$

$$A2 = \begin{pmatrix} -R1/L1 & -1/L1 \\ 1/C1 & -1/(R2 * C1) \end{pmatrix};$$

B1=(E1/L1,0);

$$X0 = \begin{pmatrix} 0 \\ 0 \end{pmatrix};$$

In the next cell the matrix exponents $e^{A_1 t_1}$ and $e^{A_2 t_2}$ are calculated (where $t_2 = T - t_1$).

At1=MatrixExp[A1*t1];

At2=MatrixExp[A2*t2];

AT=At2.At1;

A1inv=Inverse[A1];

A2inv=Inverse[A2];

I2=IdentityMatrix[2];

ATinv=Inverse[I2-AT];

XT=ATinv.(At2.A1inv.(At1-I2)+A2inv.(At2-I2)).B1

Xt1=At1.XT+A1inv.(At1-I2).B1

The inverse matrices A_1^{-1} and A_2^{-1} (denoted **A1inv, A2inv**) are calculated by the **Inverse[ ]** function. The 2×2 unit matrix I (denoted **I2**) is defined by the **IdentityMatrix[2]** function. The inverse matrix $(I - e^{A_2(T-t_1)}e^{A_1t_1})^{-1}$ is denoted by the symbol **Atinv**. In the rows

XT=ATinv.(At2.A1inv.(At1-I2)+A2inv.(At2-I2)).B1

Xt1=At1.XT+A1inv.(At1-I2).B1

the initial conditions $X(nT)$ and $X(nT+t_1)$ are calculated for steady-state behavior. The variable **XT** corresponds to the Equation 2.15, and the variable **Xtl** corresponds to the first equation of the set (2.14).

In order to draw a steady-state process, it is necessary to define a function that joins solutions of the set (2.13). In the row

Y1[t_]:=If[Floor[t/T]*T<t<=Floor[t/T]*T+t1,MatrixExp[A1*(t-Floor [t/T]*T)].XT+

A1inv.(MatrixExp[A1*(t-Floor[t/T]*T)]-I2).B1,MatrixExp[A2*(t-t1-Floor[t/T]*T)].

Xt1+A2inv.(MatrixExp[A2*(t-t1-Floor[t/T]*T)]-I2).B1]

such a function is defined with the help of the **If[ ]** function. The expression

Floor[t/T]*T<t<=Floor[t/T]*T+t1

corresponds to the condition $nT \le t \le nT+t_1$. When the condition holds, the first expression (this is situated between commas in the **If[, ,]** function) is equal to the **Yl[t_]** function; otherwise, the second one is equal to this function (this is situated after the second comma in the **If[, ,]** function). It should be pointed out once more that the use of the symbol ':=' shows that the calculation and assignment are executed only when the function **Y1[ ]** is invoked. The **Floor[ ]** function calculates the integer part of an argument.

The plotting of the steady-state process for the current is determined by the function

Plot[Part[Y1[t],1],{t,0,2*T},AxesLabel->{"t","i"},TextStyle->

{FontFamily->"Times",FontSize->12},GridLines->Automatic];

The mesh is outputted by the option **GridLines->Automatic**, and the font and its size are determined by the option **TextStyle->{FontFamily->"Times",FontSize -> 12}**. The graph of the steady-state process for the current is presented in Figure 2.11.

The graph for the steady-state process of the voltage is plotted similarly:

Plot[Part[Y1[t],2],{t,0,2*T},AxesLabel->{"t","u"},TextStyle->

{FontFamily->"Times",FontSize->12},GridLines->Automatic];

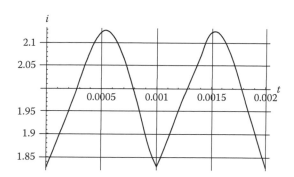

FIGURE 2.11
Steady-state process of the current (i in amperes, time t in seconds).

The time diagram is presented in Figure 2.12.

The calculation of the transitional process is realized on the basis of the recurrent use of the expressions (2.13) for the given initial condition $X(0)$. First, we calculate the transitional process in the points $X(nT)$ and $X(nT + t_1)$ using the expression (2.14).

In the row

$$\textbf{Xn1[1]=X0;}$$

the initial condition $X(nT)$ for $n = 0$ is given. The value of the variable **Kper=8** inputted in the next row of this cell defines the number of periods on which the transitional process is calculated. In the row

$$\textbf{Xn2[1]=At1.X0+A1inv.(At1-I2).Ev;}$$

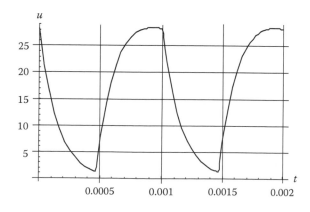

FIGURE 2.12
Steady-state process of the voltage (u in volts, time t in seconds).

the initial condition $X(nT + t_1)$ for $n = 0$ is calculated. With the help of the function

For[k=1,k<=Kper,t0=k*T;Xn2[k]=At1.Xn1[k]+A1inv.(At1-I2).B1;

Xn1[k+1]=At2.Xn2[k]+A2inv.(At2-I2).B1;k++];

the calculations of the initial conditions **Xn1[k]** and **Xn2[k]** for **Kper** periods are produced. Second, we define the function that describes the transitional process for an arbitrary period, taking into account the initial conditions

Y2[t_]:=If[Floor[t/T]*T<t<=Floor[t/T]*T+t1,MatrixExp[A1*

(t-Floor[t/T]*T)].Xn1[Floor[t/T]+1]+A1inv.(MatrixExp[A1*

(t-Floor[t/T]*T)]-I2).B1,MatrixExp[A2*(t-t1-Floor[t/T]*T)].

Xn2[Floor[t/T]+1]+A2inv.(MatrixExp[A2*(t-t1-Floor[t/T]*T)]-I2).B1];

Time diagrams of the transition processes for the current and voltage are generated with the help of the function

Plot[Part[Y2[t], 1], {t, 0, Kper*T}, AxesLabel -> {"t", "i"}, TextStyle -> {FontFamily -> "Times", FontSize -> 12}, GridLines -> Automatic];

and correspondingly by the function

Plot[Part[Y2[t], 2], {t, 0, Kper*T}, AxesLabel -> {"t", "u"}, TextStyle -> {FontFamily -> "Times", FontSize -> 12}, GridLines -> Automatic];

Time diagrams of the processes are shown in Figures 2.13 and 2.14. Analysis of the processes represented in Figures 2.13 and 2.14 show that

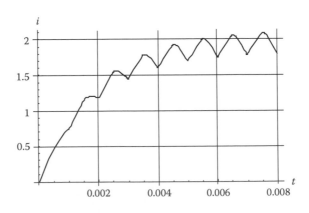

FIGURE 2.13
Transition process of the current (*i* in amperes, time *t* in seconds).

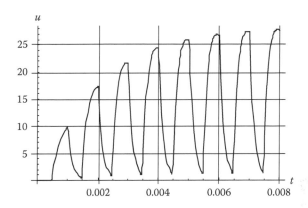

FIGURE 2.14
Transition process of the voltage (u in volts, time t in seconds).

the transient behavior is finished approximately through eight periods of the generator voltage.

2.4 Calculation of Processes in Nonlinear Systems

Let us consider the calculation of the transition process in the circuit of the noncontrolled rectifier (Figure 2.15)

The equations that describe changes of the voltage u on the capacitor, the current i, and the voltage on a diode u_d are given by

$$e(t) = u_d + u;$$

$$i = C\frac{du}{dt} + \frac{u}{R};$$ (2.16)

$$i = f(u_d),$$

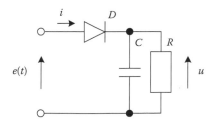

FIGURE 2.15
Circuit of the noncontrolled rectifier.

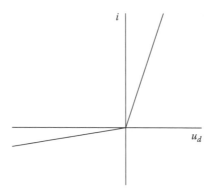

FIGURE 2.16
Approximated voltage-current characteristic of the diode.

where $i = f(u_d)$ is the voltage-current characteristic of the diode. We present this characteristic in a view of two sections of a straight line (Figure 2.16). The mathematical description of such characteristic has the form

$$R_d = \begin{cases} R_{d1}, & u_d \geq 0, \\ R_{d2}, & u_d < 0 \end{cases} \tag{2.17}$$

Solving the set of equations (2.16), we obtain the nonlinear differential equation with respect to the voltage across the capacitor:

$$\frac{du}{dt} = \frac{1}{CR_d}e(t) - \frac{1}{C}\left(\frac{1}{R_d} + \frac{1}{R}\right)u.$$

Let us consider how to solve this equation by means of Mathematica. In the first row of the cell, the use of the function

Clear[sol];

allows the cleaning of the **sol** variable. This is necessary in the case when a repeated calculation takes place (for example, after one or several parameter changes).

In the next rows the variables are defined and their values are assigned:

Rd1=0.1;

Rd2=20000.0;

R=10.0;

C1=1000.0*10^(-6);

f=50.0;

ω=2*π*f;

In the row

Rdi[ud_]:=If[ud>=0,Rd1,Rd2];

the function (2.17) is determined. In the row

e[τ_]:=20.0*Sin[ω*τ];

the function corresponding to the input voltage is defined. In the row

ud:=e[t]-Part[Evaluate[u[t]/.sol],1];

the variable of the diode voltage is defined. In the expression

sol=NDSolve[{u′[t]==1/(C1*Rdi[ud])*e[t]-1/C1*
(1/Rdi[ud]+1/R)*u[t],u[0]==0},u,{t,0,0.05}]

the **sol** variable is used, to which the solution to the differential equation is assigned later on. In the output row

{{u->InterpolatingFunction[{{0.,0.05}},<>]}}

Mathematica shows that the value of the variable **u** is approximated successfully.

In the row

Plot[{Part[Evaluate[u[t]/.sol],1],ud,e[t]},{t,0.0,0.042},AxesLabel->{"t","u"},

DefaultFont->{"Arial",12}}];

plotting of the time diagrams of the voltage on the capacitor, diode, and supply voltage is produced (Figure 2.17). The analysis of the voltages on the

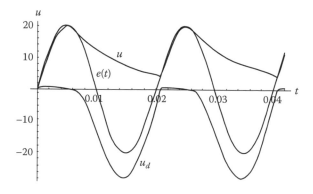

FIGURE 2.17
Time diagrams of the voltages on the diode u_d, capacitor u, and power supply $e(t)$ ([u_d, u, and $e(t)$ in volts, time t in seconds]).

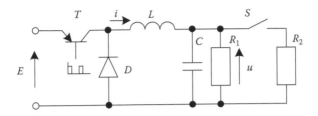

FIGURE 2.18
Circuit of the converter with periodically commutated load.

capacitor and diode shows that changes in responses of the voltages occur simultaneously.

2.5 Calculation of Processes in Systems with Several Aliquant Frequencies

In open-loop stable systems with switches, the steady-state process does not exist if the periods of switching are aliquant. Let the transistor T and switch S in the circuit of the converter presented in Figure 2.18 be switched periodically with periods T and Θ, and at such periods be aliquant.

In the area of one independent variable of time t, steady-state behavior does not exist. However, when we introduce the second independent variable of time τ, then, in the area of two variables t and τ, steady-state behavior exists (Korotyeyev, 1999). The simplest example illustrating this fact is the electric circuit (Figure 2.19) with two independent periodic power supplies.

The current in such a circuit is

$$i(t) = \frac{e_1(t) + e_2(t)}{R},$$

where $e_1(t) = e_1(t+T)$, $e_2(t) = e_2(t+\Theta)$.

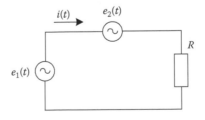

FIGURE 2.19
Circuit with two independent power supplies.

Since in such a system reactive elements are absent, the current immediately becomes quasi-periodical. Note that $i(t) \neq i(t+T)$ and $i(t) \neq i(t+\Theta)$. In the area of the two independent variables t and τ, steady-state behavior exists. Let us define the current

$$i(t, \tau) = \frac{e_1(t) + e_2(\tau)}{R}$$

Then the current $i(t, \tau) = i(t+T, \tau+\Theta)$ is periodical.

In electrical circuits with reactive elements, the introduction of the additional independent variable causes the necessity for a change of differential equations. When the power suppliers $e_1(t)$ and $e_2(t)$ work on the RL-load, the method of superposition can be used for the computation of quasi-steady-state processes. According to this method, when power supply $e_2(t)$ is shorted, the process is described by the differential equation

$$L\frac{di(t)}{dt} + Ri(t) = e_1(t),\tag{2.18}$$

When power supply $e_1(t)$ is shorted, the process is described by the differential equation

$$L\frac{di(\tau)}{d\tau} + Ri(\tau) = e_2(\tau),\tag{2.19}$$

We define the current as follows:

$$i(t, \tau) = i(t) + i(\tau).$$

Then, summing the right and left parts of the Equations 2.18 and 2.19, we can write the process in such a circuit by the differential equation

$$L\frac{\partial i(t, \tau)}{\partial t} + L\frac{\partial i(t, \tau)}{\partial \tau} + Ri(t, \tau) = e_1(t) + e_2(\tau).$$

In what follows, this reasoning will form the basis of a model expansion during the analysis of electromagnetic processes in converters.

Let us consider the calculation of the processes in the converter (Figure 2.18), with the same assumptions for active and passive elements. Then, electromagnetic processes are described by the nonstationary differential equation

$$\frac{dX(t)}{dt} = A(t)X(t) + B(t),\tag{2.20}$$

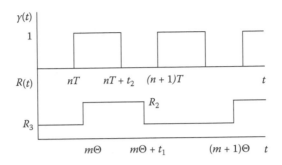

FIGURE 2.20
Time diagrams of the functions $R(t)$ and $\gamma(t)$.

where

$$X(t) = \begin{vmatrix} i(t) \\ u(t) \end{vmatrix} \; ; \quad A(t) = \begin{vmatrix} -\dfrac{r}{L} & -\dfrac{1}{L} \\[2mm] \dfrac{1}{C} & -\dfrac{1}{CR(t)} \end{vmatrix} \; ; \quad B(t) = \begin{vmatrix} \dfrac{E\gamma(t)}{L} \\[2mm] 0 \end{vmatrix} \; ;$$

the functions $R(t)$ and $\gamma(t)$ are shown in Figure 2.20; $R_3 = \frac{R_1 R_2}{R_1 + R_2}$. The matrix $A(t) = A(t+\Theta)$ and vector $B(t) = B(t+T)$ are periodical; moreover, the periods T and Θ are aliquant.

Using the Lyapunov transformation (Gantmacher, 1977)

$$X(t) = F(t)Y(t) \tag{2.21}$$

we transform the differential equation with periodical coefficients into the differential equation with constant coefficients:

$$\frac{dY(t)}{dt} = KY(t) + N(t)B(t), \tag{2.22}$$

where $F(t) = F(t+\Theta)$ is Lyapunov's matrix; $Y(t)$ is the new vector of state variables; and $N(t)$ is the inverse matrix for the matrix $F(t)$.

Matrices $F(t)$ and K are defined by the equation

$$\frac{dF(t)}{dt} = A(t)F(t) - F(t)K \tag{2.23}$$

and the conditions $F(t) = F(t+\Theta)$, $F(0) = I$ (I being the unit matrix).

Let us solve Equation 2.23 on the intervals of the matrix's $A(t)$ constancy. On the interval $m\Theta \le t \le m\Theta + t_1$, Equation 2.23 takes the form

$$\frac{dF(t)}{dt} = A_1 F(t) - F(t)K, \tag{2.24}$$

where $A_1 = A(t)$ at $R(t) = R_1$. The solution to Equation 2.24 is (Bellman, 1976)

$$F(t) = e^{A_{1t}} F(0) e^{-Kt}. \tag{2.25}$$

Similarly, for the interval $m\Theta + t_1 \le t \le (m+1)\Theta$, the solution to Equation 2.23 is

$$F(t) = e^{A_2(t-t_1)} F(t_1) e^{-K(t-t_1)}, \tag{2.26}$$

where $A_2 = A(t)$ at $R(t) = R_3$.

Substituting $t = t_1$ in (2.25), $t = \Theta$ in (2.26), and then eliminating $F(t_1)$ from the obtained expressions, the following is obtained:

$$F(0) = e^{A_2(\Theta - t_1)} e^{A_{1t_1}} e^{-K\Theta}. \tag{2.27}$$

Taking into account that $F(\Theta) = F(0) = I$, we find the matrix from (2.27):

$$K = \frac{1}{\Theta} \ln[e^{A_2(\Theta - t_1)} e^{A_{1t_1}}]. \tag{2.28}$$

Then, for the interval $m\Theta \le t \le m\Theta + t_1$ the matrix $F(t)$ is

$$F(t) = e^{A_{1t}} e^{-Kt}, \tag{2.29}$$

and for the interval $m\Theta + t_1 \le t \le (m+1)\Theta$, the matrix $F(t)$ is

$$F(t) = e^{A_2(t-t_1)} e^{A_{1t_1}} e^{-Kt}. \tag{2.30}$$

Similar to the given reasoning about the model expansion for the two power supplies, we introduce one more independent variable of time τ and expand Equation 2.22 in the following way:

$$\frac{\partial Y(t, \tau)}{\partial t} + \frac{\partial Y(t, \tau)}{\partial \tau} = KY(t, \tau) + N(t)B(\tau). \tag{2.31}$$

To define the steady-state process we apply the multidimensional Laplace transform (Pupkov et al., 1976) to Equation 2.31. Then,

$$[(p+q)I - K]Y(p, q) = N(p)B(q), \tag{2.32}$$

where p,q are the complex variables of the multidimensional Laplace transform; $Y(p,q)$, $N(p)$, and $B(q)$ are the Laplace transforms of the functions $Y(t,\tau)$, $N(t)$, and $B(\tau)$.

The solution to Equation 2.32 has the form

$$Y(p,q) = W(p,q)N(p)B(q), \tag{2.33}$$

where $W(p,q) = [(p+q)I - K]^{-1}$.

Let us transform (2.21) into the expression of the two independent variables

$$X(t,\tau) = F(t)Y(t,\tau)$$

and then apply the multidimensional Laplace transform to this expression. We obtain

$$X(p,q) = F(p) * Y(p,q), \tag{2.34}$$

where $*$ is the sign of convolution in the p–q domain. Since the matrices $F(t)$ and $N(t)$, and the vector $B(t)$ are periodical, their transformations have the forms

$$F(p) = \frac{F_\Theta(p)}{1-e^{-p\Theta}}, \quad N(p) = \frac{N_\Theta(p)}{1-e^{-p\Theta}}, \quad B(q) = \frac{B_T(q)}{1-e^{-qT}},$$

where

$$F_\Theta(p) = \int_0^\Theta e^{-pt}F(t)\,dt, \ N_\Theta(p) = \int_0^\Theta e^{-pt}N(t)\,dt, \ B_T(q) = \int_0^T e^{-qt}B(\tau)\,d\tau.$$

Let us find the convolution in the expression (2.34). Since the convolution with respect to poles of the function $F_\Theta(p)$ gives zero value, and the poles of this function do not coincide with the poles of the function $\frac{1}{1-e^{-p\Theta}}$ (the considering circuit is dissipative), we find the convolution with respect to the poles of the function $\frac{1}{1-e^{-p\Theta}}$. Then,

$$X(p,q) = \frac{1}{\Theta}\sum_{k=-\infty}^{\infty} F_\Theta(p_k)W(p-p_k,q)N(p-p_k)B(q), \tag{2.35}$$

where p_k are the roots of the equation $1-e^{-p\Theta} = 0$; $p_k = j\frac{2\pi k}{\Theta}$ $(k = 0,\pm 1,\pm 2,\ldots)$

Let us present the steady-state process in the form of an aliquot Fourier series (Tolstoy 1951). Transformation of vector $B(t)$ is

$$B(q) = \begin{vmatrix} \dfrac{E}{L}\dfrac{1-e^{-qt_2}}{q(1-e^{-qT})} \\ 0 \end{vmatrix} = \dfrac{E}{L}\begin{vmatrix} \dfrac{1-e^{-qt_2}}{q} \\ 0 \end{vmatrix}\begin{vmatrix} \dfrac{1}{1-e^{-qT}} \end{vmatrix}.$$

Then, the inverse Laplace transform for (2.35), which is calculated with respect to the poles of the functions $N(p-p_k)\,\frac{1}{1-e^{-p\Theta}}$ and $B(q)\frac{1}{1-e^{-qT}}$, gives

$$X(t,\tau) = \dfrac{1}{\Theta^2 T}\sum_{\substack{m,n=-\infty \\ n\neq 0}}^{\infty}\left[\sum_{k=-\infty}^{\infty}F_\Theta(p_k)W(p_m-p_k,q_n)N_\Theta(p_m-p_k)B_T(q_n)\right]e^{p_m t}e^{q_n \tau} + X(t,0),$$

$$(2.36)$$

where $p_m = j\frac{2\pi m}{\Theta}$ $(m=0,\pm 1,\pm 2,...)$ are the roots of the equation $1-e^{-p\Theta} = 0$; $q_n = j\frac{2\pi n}{T}$ $(n=0,\pm 1,\pm 2,...)$ are the roots of the equation $1-e^{-qT}=0$; and $B_T(q_n) = \begin{vmatrix} \frac{E(1-e^{-q n t_2})}{TLq_n} \\ 0 \end{vmatrix}$. In the expression (2.36), the second term is given by

$$X(t,0) = \dfrac{1}{\Theta^2}\sum_{m=-\infty}^{\infty}\left[\sum_{k=-\infty}^{\infty}F_\Theta(p_k)W(p_m-p_k,0)N_\Theta(p_m-p_k)B_T'(0)\right]e^{-p_m t}, \quad (2.37)$$

where

$$B_T'(0) = \lim_{q\to 0} qB(q) = \begin{vmatrix} E\dfrac{t_1}{T} \\ 0 \end{vmatrix}.$$

Let us consider how to find the quasi-steady-state values of the current $i(t)$ and voltage $u(t)$ with the help of Mathematica. In the first cell we enter the parameters of the circuit elements:

Rs=1.6;

L1=0.2*10^(-3);

$$C1=10*10^{\wedge}(-6);$$

$$R1=10.0;$$

$$R3=5.0;$$

$$A1 = \begin{pmatrix} -Rs/L1 & -1/L1 \\ 1/C1 & -1/(R3*C1) \end{pmatrix};$$

$$A2 = \begin{pmatrix} -Rs/L1 & -1/L1 \\ 1/C1 & -1/(R1*C1) \end{pmatrix};$$

$$t1=4*10^{\wedge}(-5);$$

$$\Theta = 6*10^{\wedge}(-5);$$

$$t2=8.0*10^{\wedge}(-5);$$

$$T=10.0*10^{\wedge}(-5);$$

$$E1=12.0;$$

$$Ns=2;$$

$$K\Theta = 2*Pi/\Theta;$$

$$KT=2*Pi/T;$$

I2=IdentityMatrix[Ns];

In this cell, Rs denotes *r*, **KΘ** defines the angular frequency for the period Θ, **KT** defines the angular frequency for the period *T*, and Ns defines the order of the matrix $A(t)$.

In the following cell the calculation of the matrix K is produced according to the expression (2.28). Since Mathematica does not have a built-in function for the matrix logarithm calculation, an integral calculation is used in the program. For matrix A the logarithm is calculated in the following way (Davies and Higham, 2005):

$$\ln[A] = \int_0^1 (A-I)[x(A-I)+I]^{-1}dx. \tag{2.38}$$

This expression is true if the matrix argument does not have eigenvalues on the negative part of the real axis.

The matrix K is signified as K1 in the program. The matrix logarithm is calculated in the cell

A21=MatrixExp[A2*(Θ-t1)].MatrixExp[A1*t1];

K1=Integrate[(A21-I2).Inverse[x*(A21-I2)-I2]/ Θ,{x,0,1}];

In the following two cells the calculation of the $F_\Theta(p)$ function is produced. The function is calculated for the intervals of the matrix $F(t)$ constancy. The parts of the $F_\Theta(p)$ function are denoted as **F1** and **F2** in the program.

Clear[pk]

Fnt1=MatrixExp[A1*t].MatrixExp[-K1*t];

$$F1 = \int_0^{t1} Fnt1 * E\wedge(-pk * t)dt;$$

Fnt2=MatrixExp[A2*t].MatrixExp[A2*(-t1)].MatrixExp[A1*t1].MatrixExp[-(K1*t)];

$$F2 = Simplify\left[\int_{t1} Fnt2 * E\wedge(-pk * t)dt \right];$$

In these expressions, **pk** is an independent variable, for which a value is set later.

The $N_\Theta(p)$ function is calculated similarly. Since $N(t) = F^{-1}(t)$, when the integrals are calculated, the inverse matrices are found initially for the intervals of constant topology of the matrix $F(t)$. At those for nonsingular matrices A and B, we use $(AB)^{-1} = B^{-1}A^{-1}$. Then, on the intervals of the matrix $F(t)$ constancy, we find matrices **N1** and **N2**:

NInt1=MatrixExp[K1*t].MatrixExp[-A1*t];

Clear[p]

$$N1 = \int_0^{t1} NInt1 * E\wedge(-p * t)dt$$

NInt2=MatrixExp[K1*t].MatrixExp[-A1*t1].MatrixExp[A2*t1].MatrixExp[-A2*t];

$$N2 = Simplify\left[\int_{t1} NInt2 * E\wedge(-p * t)dt \right];$$

In these expressions, **p** is an independent variable, for which a value is set later.

The inverse matrix

$$W(p,q) = [(p-q)I - K]^{-1}$$

is formed in the following way (the matrix is designated **WK** in the program):

Clear[pq,n,m]

KK=(pq-pk)*I2-K1;

WK=Inverse[KK];

pk=I*k*KΘ;

qn=I*n*KT;

pm=I*m*KΘ;

pq=pm+qn;

p=pm-pk;

The variables **pk**, **qn**, **pm**, and **p** are also defined in the cell.

In the next cell the constants **num**, **knum**, **kn**, and a matrix **X1** are defined. The **knum** constant defines the number of summands in a $\sum_{k=-knum}^{knum}$ sum, the **num** constant defines the number of summands in the $\sum_{m,n=-num}^{num}$ sum, and the **kn=2*num+1** constant defines the dimension of the matrix **X1**. In the program, this matrix is defined by the expression

$$\sum_{k=-knum}^{knum} F_\Theta W(p_m - p_k, q_n) N_\Theta(p_m - p_k) B_T(q_n)$$

which corresponds to the part of (2.36):

num=3;

knum=10;

kn=2*num+1;

Array[X1,{kn,kn}];

Bn=E1*(1-E^(-qn*t2))/(qn*T*L1*Θ^2);

Z0={{1.*10^(-8),1.*10^(-8)},{1.*10^(-8),1.*10^(-8)}};

For[m=-num, m<num+1, m++,For[n=-num, n<num+1, n++, If[n==0, X1[m+1+num,n+1+num]=Z0, X1[m+1+num,n+1+num]=Bn*Sum[(F1+F2). WK.(N1+N2),{k,-knum,knum,1}]]]]]

By means of the **Array[]** function, the matrix **X1**, elements of which are matrices, is created.

In the program, the first element of the vector $B_T(q_n)$, together with overall coefficient $1/\Theta^2$ in (2.36), is defined as

Bn=E1*(1-E^(-qn*t2))/(qn*T*L1*Θ^2)

All the matrix elements are multiplied by this coefficient. The second element of the vector $B_T(q_n)$ is equal to zero. Therefore, the corresponding elements of the matrix **X1** are not used later on.

The matrix **Z0** is entered to fill the matrix **X1** for $n = 0$. In the expression (2.36), these values are not defined (summing is produced for $n \neq 0$). To simplify the following calculations, it could be taken into account that **Z0** is equal to the zero matrix. However, when the coefficients of a Fourier series are found, the calculation of the arguments of the matrix elements is produced. This leads to infinity for some coefficients. To simplify the transformations we define the matrix **Z0** as sufficiently small. We calculate the matrix **X1** with the help of the **For[]** function.

In the next cell the variables **pm**, **pq**, and **p**, and the coefficient $B_T'(0)$ and a matrix **X01** are defined:

$$\text{Array[X01,kn];}$$

$$\text{Clear[m]}$$

$$\text{pm=m*I*K}\Theta;$$

$$\text{pq=pm;}$$

$$\text{p=pm-pk;}$$

$$\text{B0=E1*t2/(T*L1*}\Theta\text{\textasciicircum2);}$$

$$\text{For[m=-num,m<num+1,X01[m+1+num]=Sum[(F1+F2).WK.(N1+N2),}$$

$$\text{\{k,knum,knum,1\}]*B0;m++];}$$

The matrix **X01**, the elements of which are the matrices of (2.37)

$$\sum_{k=-knum}^{knum} F_\Theta(p_k)W(p_m - p_k, 0)N_\Theta(p_m - p_k)B_T'(0)$$

is created by means of the **Array[]** function. The first element of the vector $B_T'(0)$ and the overall coefficient $1/\Theta^2$ in the program are defined as **B0**. As in the previous case, all matrix elements are multiplied by this coefficient. However, further, during the calculation, only those elements are used that correspond to the nonzero values.

The expression (2.36) for the current **It[th_]** of the one-variable function $t = \tau$ is formed in the cell

$$\text{It[th\_]:=Re[Sum[Sum[Part[X1[m+num+1,n+num+1],1,1]*}$$

$$\text{E\textasciicircum(I*K}\Theta\text{*m*th+I*KT*n*th),\{n,-num,num\}],\{m,-num,}$$

$$\text{num\}]]+Re[Sum[Part[X01[k+1+num],1,1]*E\textasciicircum(I*K}\Theta\text{*k*th),\{k,-num,num\}]]}$$

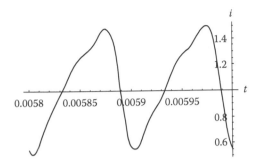

FIGURE 2.21
The quasi-steady-state process of the current for $0.0058 \le t \le 0.006$ (i in amperes, time t in seconds).

The required number of terms of the series is calculated with the help of the **Sum[]** function. By means of the **Re[]** function, the terms with imagined parts, which arise through inaccuracies and summing of the complex expressions, are removed.

The time diagram for the current is plotted with the help of the function

Plot[It[th],{th,0.006-2*T,0.006},AxesLabel->{"t","i"}]

The quasi-steady-state process of the current is shown in Figure 2.21.

It can be seen from the figure that the current function for such parameters is practically periodical on the interval $0-2T$ of time.

The expression **Ut[th_]** for the voltage of the one-variable function $t = \tau$ is composed similarly:

Ut[th_]:=Re[Sum[Sum[Part[X1[m+num+1,n+num+1],2,1]*E^(I*KΘ*m*th+

I*KT*n*th),{n,-num,num}],{m,-num,num}]]+
Re[Sum[Part[X01[k+num+1],2,1]*

E^(I*KΘ*k*th),{k,-num,num}]]

The graph of the voltage (represented in Figure 2.22) is plotted with the help of the function **Plot[Ut[th],{th,0,2*T},AxesLabel->{"t","u"}]**. It can be seen from the figure that the voltage is not a periodical function.

The expression **Itτ[ta_,tc_]** for the current of the two-variable function t and τ is formed in the following way:

Itτ[ta_,tc_]:=Re[Sum[Sum[Part[X1[m+num+1,n+num+1],1,1]*E^

(I*KΘ*m*ta+I*KT*n*tc),{n,-num,num}],{m,-num,num}]]+

Re[Sum[Part[X01[k+1+num],1,1]*E^(I*KΘ*k*ta),{k,-num,num}]]

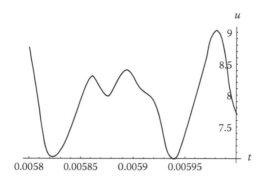

FIGURE 2.22
The quasi-steady-state process of the voltage for $0.0058 \leq t \leq 0.006$ (u in volts, time t in seconds).

The graph of the steady-state process of the current is generated by means of the function

Plot3D[Itτ[ta,tc],{ta,0,Θ},{tc,0,T},Ticks->{{0,0.00003,0.00006},
{0,0.00005,0.0001},

{0,1}},Lighting->False,AxesLabel->{"t","τ","i"}]

and is presented in Figure 2.23. With the help of the option **Ticks->**
{{0,0.00003,0.00006}, {0,0.00005,0.0001},{0,1}}, the required tick marks on the axes are set explicitly.

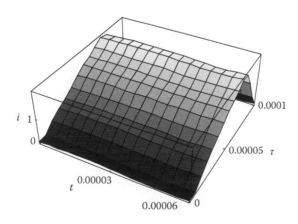

FIGURE 2.23
The steady-state process of the current (i in amperes, t and τ in seconds).

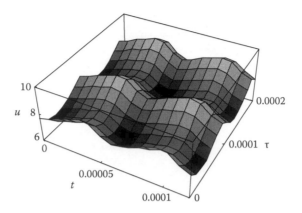

FIGURE 2.24
The steady-state process of the voltage (u in volts, t and τ in seconds).

The expression **Utτ[ta_,tc_]** for the voltage of the two-variable function t and τ is formed similarly:

$$\textbf{Ut}\tau\textbf{[ta\_,tc\_]:=Re[Sum[Sum[Part[X1[m+num+1,n+num+1],2,1]*}$$
$$\textbf{E\^{}(I*K}\Theta\textbf{*m*ta+}$$

$$\textbf{I*KT*n*tc),\{n,-num,num\}],\{m,-num,num\}]]+Re[Sum[Part[X01}$$
$$\textbf{[k+1+num],2,1]*}$$

$$\textbf{E\^{}(I*K}\Theta\textbf{*k*ta),\{k,-num,num\}]]}$$

The steady-state process for the two periods of the voltage is generated by means of the function

$$\textbf{Plot3D[Ut}\tau\textbf{[ta,tc],\{ta,0,2*}\Theta\textbf{\},\{tc,0,2*T\},Ticks->\{\{0,0.00005,0.0001\},}$$
$$\textbf{\{0,0.00001,0.0002\}, \{6,8,10\}\},Lighting->False,PlotRange->\{6,10\},}$$
$$\textbf{AxesLabel->\{"t","}\tau\textbf{","u"\}]}$$

and is represented in Figure 2.24. The plotting of a black-and-white picture is realized by means of the option **Lighting -> False.** The option **PlotRange -> {6,10}** provides the range of outputted values.

2.6 Analysis of Harmonic Distribution in an AC Voltage Converter

Let us consider an analysis of the harmonic distribution and steady-state process calculation in a system with an AC converter (Figure 2.25).

Suppose that a period of topology change of a converter as well as a period of an input voltage are aliquant. To find the steady-state process in such a system in which a period of the supplying voltage and a period of the switching of a

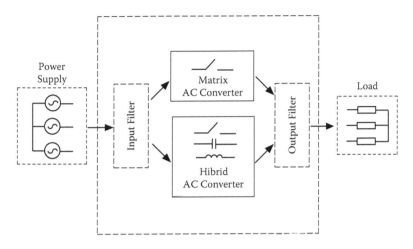

FIGURE 2.25
AC converter in a power supply system. (Data from Korotyeyev I. Ye., Fedyczak Z. Analysis of steady-state behavior in converters with changed topology Technical electrodynamics, Supply System of Electrotechnical Devices and Systems, Kiev, No. 1, pp. 31–34, 1999).

converter are aliquant, it is necessary to expand the initial area of one variable to the area of several independent variables of time. The expansion is realized by the substitution of the periodical functions, which correspond to the independent periodical signals, for the functions with independent variables of time. In addition, the derivatives of one independent variable are substituted for the sum of derivatives of all independent variables. With such expansion the steady-state process exists in the area of several independent variables.

For an AC converter we use the Boost converter (Figure 2.26).

Let us consider that the power switches S_1 and S_2 are bidirectional. Furthermore, if the key S_1 is opened, then the key S_2 is closed, and vice versa. Suppose that the switches are described by the RS model and have the same resistance in the on state. The electromagnetic processes in such systems are described by the nonstationary matrix differential equation (2.20) in which

$$X(t) = \begin{vmatrix} i(t) \\ u(t) \end{vmatrix}$$

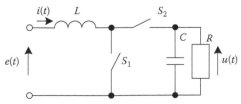

FIGURE 2.26
AC Boost converter.

is the vector of the state variables,

$$A(t) = \begin{vmatrix} -\dfrac{r}{L} & -\dfrac{\gamma(t)}{L} \\ \dfrac{\gamma(t)}{C} & -\dfrac{1}{RC} \end{vmatrix}, \quad B(t) = E'e(t), \quad E' = \begin{vmatrix} \dfrac{1}{L} \\ 0 \end{vmatrix};$$

$r = r_s + r_L$ is the sum of resistances of the closed switch r_s and the inductor r_L, $e(t) = USin\omega t$, $\omega = 2\pi/T$, T is the period of the supplying voltage, and Θ is the switching period of power switches. The state of the switches is described by the switching function $\gamma(t)$ (Figure 2.27). The off state of the key S_1 and the on state of the key S_2 correspond to the zero value of the switching function. When $\gamma(t) = 1$, key S_1 is opened and key S_2 is closed.

Using the Lyapunov transformation (2.21), $X(t) = F(t)Y(t)$, and expanding the area of one independent variable of time t to the area of two independent variables t and τ (Korotyeyev and Fedyczak, 1999), let us present the nonstationary equation in the form (2.31):

$$\frac{\partial Y(t,\tau)}{\partial t} + \frac{\partial Y(t,\tau)}{\partial \tau} = KY(t,\tau) + N(t)B(\tau), \tag{2.39}$$

Applying the multidimensional Laplace transform to Equation 2.39, we find the solution (2.34), which can be represented in the following way:

$$X(p,q) = F(p) * [W(p,q)N(p)B(q)], \tag{2.40}$$

where $F(p) = \frac{F_\Theta(p)}{1-e^{-p\Theta}}$, $N(p) = \frac{N_\Theta(p)}{1-e^{-p\Theta}}$, $B(q) = E'\frac{U\omega}{q^2+\omega^2}$, $W(p,q) = [(p+q)I - K]^{-1}$.

Calculating the convolution in (2.40) with respect to the poles of the $F(p)$ function yields (2.35) in the form

$$X(p,q) = \frac{1}{\Theta(1-e^{-p\Theta})} \sum_{k=-\infty}^{\infty} F_\Theta(p_k)W(p-p_k,q)N_\Theta(p-p_k)B(q). \tag{2.41}$$

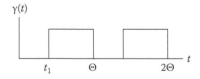

FIGURE 2.27
Time diagram of the switching function.

Let us present the steady-state process for $X(p,q)$ as the double Fourier series (Tolstoy, 1951):

$$X_s(t,\tau) = \sum_{m,n=-\infty}^{\infty} C_{m,n} e^{j(m\vartheta t + n\omega\tau)}, \qquad (2.42)$$

where $\vartheta = \frac{2\pi}{\Theta}$. The double Fourier series is obtained as the inverse Laplace transform for the expression (2.41) with respect to the poles of the $N(p)$ function $\frac{1}{1-e^{-p\Theta}}$, which correspond to the steady-state process and to the poles of the function $B(q)$, that is, $q_{1,2} = \pm j\omega$. Taking into account this reasoning, the expression (2.42) takes the form

$$X_s(t,\tau) = \sum_{m=-\infty}^{\infty} \sum_{\substack{n=-1 \\ n \neq 0}}^{1} C_{m,n} e^{j(m\vartheta t + n\omega\tau)} \qquad (2.43)$$

where

$$C_{m,n} = \frac{nU}{j2\Theta^2} \sum_{k=-\infty}^{\infty} F_\Theta(p_k)W(p_m - p_k, q_n)N_\Theta(p_m - p_k)E' \qquad (2.44)$$

$p_m = j\frac{2\pi m}{\Theta}$ are the roots of the equation $1 - e^{-p\Theta} = 0$; $m = 0, \pm 1, \pm 2...$; $q_n = j\omega n$, $n = -1, 1$.

Let us consider how to use Mathematica for the calculation of the process in the system with the Boost converter. The initial data is presented in the cell

$$r1=0.2;$$

$$L1=0.15*10^\wedge(-3);$$

$$C1=60*10^\wedge(-6);$$

$$R11=0.8*10^\wedge3;$$

$$A1 = \begin{pmatrix} -r1/L1 & 0 \\ 0 & -1/(R11*C1) \end{pmatrix};$$

$$A2 = \begin{pmatrix} -r1/L1 & -1/L1 \\ 1/C1 & -1/(R11*C1) \end{pmatrix};$$

$$t1=7/5*10^\wedge(-4);$$

$$T=20*10^\wedge(-3);$$

$$\Theta=7*10^\wedge(-4);$$

$$K\Theta =2*Pi/\Theta;$$

$$KT=2*Pi/T;$$

$$t2=\Theta-t1;$$

$$U=310;$$

$$Ns=2;$$

$$I2=IdentityMatrix[Ns];$$

In this cell, **r1** denotes r, **R11** denotes R, **K⊖** defines the angular frequency for the period ⊖, **KT** defines the angular frequency for the period T, and Ns defines the order of matrix $A(t)$. In the following cells we calculate the matrix **K**:

$$A21 = MatrixExp[A2*t2].MatrixExp[A1*t1];$$

$$K1 = Integrate[(A21 - I2).Inverse[x*(A21 - I2) + I2]/\Theta, \{x, 0, 1\}];$$

and matrices **F1, F2, N1, N2**

$$Clear[pk];$$

$$Fnt1=MatrixExp[A1*t].MatrixExp[-K1*t];$$

$$F1 = Simplify\left[\int_0^{t1} Fnt1 * E\wedge(-pk * t)dt\right];$$

$$Fnt2=MatrixExp[A2*t].MatrixExp[A2*(-t1)].MatrixExp[A1*t1].$$
$$MatrixExp[-(K1*t)];$$

$$F2 = Simplify\left[\int_{t1} Fnt2 * E\wedge(-pk * t)dt\right];$$

$$Nint1=MatrixExp[K1*t].MatrixExp[-A1*t];$$

$$Clear[p];$$

$$N1 = \int_0^{t1} NInt1 * E\wedge(-p * t)dt;$$

$$Nint2=MatrixExp[K1*t].MatrixExp[-A1*t1].MatrixExp[A2*t1].$$
$$MatrixExp[-A2*t];$$

$$\textbf{Clear[p,t];}$$

$$\textbf{N2} = \int_{t1} \textbf{NInt2} * \textbf{E} \wedge (-p * t) dt;$$

Further, we define the inverse matrix $W(p_m - p_k, q_n)$ (denoted as **WK**), roots **pk, pm, q0**, dimensions and the number of terms in the expressions (2.43) and (2.44)

$$\textbf{Clear[pq,n,m,k];}$$

$$\textbf{KK=(pq-pk)*I2-K1;}$$

$$\textbf{WK=Inverse[KK];}$$

$$\textbf{pk=I*k*K}\Theta;$$

$$\textbf{q0=I*KT;}$$

$$\textbf{pm=I*m*K}\Theta;$$

$$\textbf{pq=pm+q0;}$$

$$\textbf{p=pm-pk;}$$

$$\textbf{nn=2;}$$

$$\textbf{num=4;}$$

$$\textbf{knum=10;}$$

$$\textbf{U0=U/(L1*}\Theta\textbf{\^{}2);}$$

The constant **nn** is equal to the number of roots of the transform of the sinusoidal function, the coefficient **U0** corresponds to the part of the coefficient included in (2.44), the **knum** constant defines the number of summands in the $\sum_{k=-knum}^{knum}$ sum as in (2.44), and the **num** constant defines the number of summands in the $\sum_{m,n=-num}^{num}$ sum as in (2.43).

In the cell

$$\textbf{Array[X1,\{2*num+1,nn\}];}$$

the list **X1**, which corresponds to the matrix with complex coefficients of the Fourier series, is defined

$$\textbf{For[m=-num,m}\leq\textbf{num,For[n=1,n}\leq\textbf{nn,\{pq=(2*n-3)*q0+pm;X1[m+1+num,}}$$

$$\textbf{n]=(2*n-3)/(2*I)*U0*Sum[(F1+F2).WK.(N1+N2),}$$
$$\textbf{\{k,-knum,knum,1\}]\};n++];m++];}$$

In this expression the coefficient **(2*n − 3)** defines the sign of the pole $q_n = jn\omega$.

The output of the complex coefficients of the Fourier series for the voltage is realized by means of the expression

For[m=1,m≤2*num+1,m++,For[n=1,n≤nn,n++,If[n==1,Print["n=",−1," ","
","m=",

m − num−1," ","Cu=",Part[X1[m,n],2,1]],Print["n=",1," ","
","m=",m−num−1,

" ","Cu=",Part[X1[m,n],2,1]]]]];

The coefficients of the complex Fourier series for the current are outputted similarly:

For[m=1,m≤2*num+1,m++,For[n=1,n≤nn,n++,If[n==1,Print["n=",-1," ","
","m=",

m−num−1," ","Ci=",Part[X1[m,n],1,1]],Print["n=",1," "," ","m=",m−num−1,
" ","Ci=",Part[X1[m,n],1,1]]]]];

The values of the coefficients for the voltage and current are presented in Table 2.1.

TABLE 2.1

The Values of the Coefficients for the Voltage and Current

Coefficient $C_{m,n}$	Voltage	Current
$C_{-4,-1}$	−4.514 − j0.147	−0.812 + j0.266
$C_{-4,1}$	5.246 − j1.094	−0.637 − j1.992
$C_{-3,-1}$	−8.028 + j8.525	−0.728 + j1.956
$C_{-3,1}$	7.663 − j13.81	−6.163 − j3.384
$C_{-2,-1}$	−1.178 + j29.585	5.702 + j3.788
$C_{-2,1}$	−12.19 − j47.488	−24.393 + j6.571
$C_{-1,-1}$	27.7 − j116.934	−87.98 + j7.931
$C_{-1,1}$	79.557 + j173.868	109.526 − j67.644
$C_{0,-1}$	21.309 + j141.051	0.111 + j34.561
$C_{0,1}$	21.309 − j141.051	0.111 − j34.561
$C_{1,-1}$	79.557 − j173.868	109.526 + j67.644
$C_{1,1}$	27.7 + j116.934	−87.98 − j7.931
$C_{2,-1}$	−12.19 + j47.488	−24.393 − j6.571
$C_{2,1}$	−1.178 − j29.585	5.702 − j3.788
$C_{3,-1}$	7.663 + j13.81	−6.163 + j3.384
$C_{3,1}$	−8.028 − j8.525	−0.728 − j1.956
$C_{4,-1}$	5.246 + j1.094	−0.637 + j1.992
$C_{4,1}$	−4.514 + j0.147	−0.812 − j0.266

The functions of the inverse Fourier transform are generated in the following cells, and the graphs of the steady-state processes of the voltage and current are plotted either.

Itτ[ta_,tc_]:=Re[Sum[Sum[Part[X1[m,n],1,1]*E^(I*KΘ*(m-num-1)*ta+

I*KT*(2*n-3)*tc),{n,1,nn}],{m,1,2*num+1}]];

Plot3D[Itτ[ta,tc],{ta,0,Θ},{tc,0,T},Lighting->False,AxesLabel->{"t","τ","i"}]

Utτ[ta_,tc_]:=Re[Sum[Sum[Part[X1[m,n],2,1]*E^(I*KΘ*(m-num-1)*ta+

I*KT*(2*n-3)*tc),{n,1,nn}],{m,1,2*num+1}]];

Plot3D[Utτ[ta,tc],{ta,0,Θ},{tc,0,T},Lighting->False,AxesLabel->{"t","τ","u"}]

In this cell, the variables **tc** and **ta** denote t and τ, respectively.

The graphs of the steady-state processes of the current and voltage are presented, respectively, in Figures 2.28 and 2.29.

It can be seen that the process with one time variable in that system is not steady state.

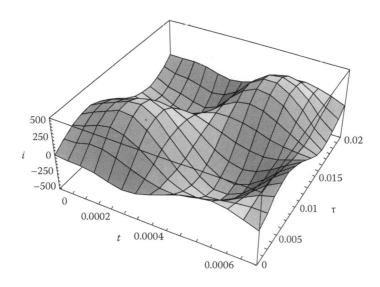

FIGURE 2.28

The steady-state process of the current time t and τ in seconds (i in amperes, t and τ in seconds).

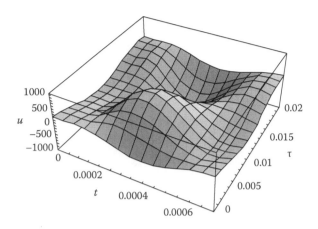

FIGURE 2.29
The steady-state process of the voltage (u in volts, t and τ in seconds).

2.7 Calculation of Processes in Direct Frequency Converter

Let us determine a steady-state current in a load on the direct frequency converter shown in Figure 2.30. Switches $S_1 - S_4$ are periodically turned on and off in such a way that the positive and negative parts of the input sinusoidal voltage, as shown in Figure 2.31, are applied to the RL load. Impulses for the switches S_1, and S_3 are shifted by half of the period from impulses for the switches S_2, and S_4.

We assume that the inductor is a linear element and the switches are ideal. Processes in this converter are described by the differential equation

$$L\frac{di(t)}{dt} + Ri(t) = u(t), \tag{2.45}$$

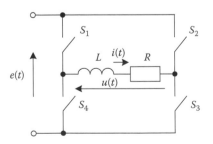

FIGURE 2.30
Topology of the direct frequency converter.

u(t)

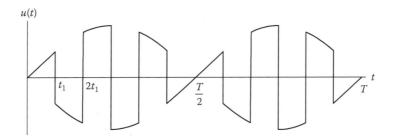

FIGURE 2.31
Voltage on the RL load.

where

$$u(t) = \begin{cases} E\sin(\omega t), & 2nt_1 \le t \le (2n+1)t_1; \\ -E\sin(\omega t), & (2n+1)t_1 \le t \le 2(n+1)t_1; \end{cases}$$

$n = 0,1,2,\ldots$, $\omega = \frac{2\pi}{T}$, T is the period of the supply voltage $e(t)$, and $t_1 = \frac{T}{2K}$ is the time interval (with this number, K must be even).

In order to determine a steady-state solution, we use the method described in Section 2.2. First we determine a Laplace transform of the voltage $u(t)$. This voltage can be obtained by multiplication of a sinusoidal function by the single rectangular pulse

$$sq(t) = \begin{cases} 1, & nt_1 \le t \le (n+1)t_1; \\ 0, & \text{otherwise}. \end{cases}$$

with the amplitude equal to one and summation of an obtained expression for $n = 0,1,\ldots,N-1$.

Let us use the convolution of two functions in the frequency domain that has the form

$$L\{f_1(t)f_2(t)\} = \frac{1}{2\pi j} \int_{c-j\infty}^{c+j\infty} F_1(s)F_2(p-s)\,ds,$$

where $L\{\ldots\}$ is the Laplace transform; $L\{f_1(t)\} = F_1(p)$; and $L\{f_2(t)\} = F_2(p)$. The complex integral can be calculated using the residue theory as follows:

$$L\{f_1(t)f_2(t)\} = \sum_{k=1}^{K_{F1}} res[F_1(s_k)F_2(p-s_k)],$$

where s_k is the k-th pole of the function $F_1(s)$; and K_{F1} is the number of poles of the function $F_1(s)$.

Since our goal is to find the steady-state solution on all intervals of period $\frac{T}{2}$, we shall derive a general expression for a function $\sin(\omega(t+t_1))$. Using Mathematica, one obtains

$$\textbf{sn=Sin[\omega*(t+t1)];}$$

$$\textbf{Lsn=LaplaceTransform[sn,t,s]}$$

$$\textbf{Lsn2=Lsn/.\{t1->n*T/K1,s->p\}}$$

$$\textbf{sgS1=((Exp[-n*s*T/K1]-Exp[-(n+1)*s*T/K1])/s)/.s->(p-s)}$$

In this cell, **sn** denotes the sinus function, **Lsn** defines the Laplace transform of the sinus function, **Lsn2** defines the Laplace transform of the sinus function for $t1 = nT/K1$ and $s = p$, **K1** denotes $2K$, **sgS1** denotes the Laplace transform of the single rectangular pulse $sq(t)$ with the substitution p for p − s. Mathematica outputs expressions

$$\frac{\omega\mathrm{Cos}\left[\frac{nT\omega}{K1}\right]+p\mathrm{Sin}\left[\frac{nT\omega}{K1}\right]}{p^2+\omega^2}$$

$$\frac{-e^{\frac{(-1-n)(p-s)T}{K1}}+e^{\frac{n(p-s)T}{K1}}}{p-s}$$

In the next cell we calculate the convolution of the two functions $\sin(\omega(t+t1))$ and $sq(t)$ for $t1 = 0$:

$$\textbf{nsnN=FullSimplify[Residue[sgS1*Lsn,\{s,I*\omega\}]]+Residue[sgS1*Lsn,}$$
$$\textbf{\{s,-I*\omega\}]]/.t1->0}$$

Mathematica outputs the expression

$$\frac{1}{2}\left(-\frac{i\left(e^{-\frac{nT(p-i\omega)}{K1}}-e^{-\frac{(1+n)T(p-i\omega)}{K1}}\right)}{p-i\omega}+\frac{i\left(e^{-\frac{nT(p+i\omega)}{K1}}-e^{-\frac{(1+n)T(p+i\omega)}{K1}}\right)}{p+i\omega}\right)$$

Then we form the Laplace transforms $\int_{t_0}^{t_0+T/2} f(t)e^{-pt}dt$ for six intervals, which is equal to the period $\frac{T}{2}$ with different initial points t_0:

$$\textbf{sg1=((nsnN/.n->0)-(nsnN/.n->1)+(nsnN/.n->2)-(nsnN/.n->3)+(nsnN/.n->4)-}$$
$$\textbf{(nsnN/.n->5));}$$

$$\textbf{sg2=-(Exp[p*T/K1]*((nsnN/.n->1)-(nsnN/.n->2)+(nsnN/.n->3)-}$$
$$\textbf{(nsnN/.n->4)+}$$

(nsnN/.n->5)+(nsnN/.n->6)));

sg3=(Exp[p*2*T/K1]*((nsnN/.n->2)-(nsnN/.n->3)+(nsnN/.n->4)
-(nsnN/.n->5)-

(nsnN/.n->6)+(nsnN/.n->7)));

sg4=-(Exp[p*3*T/K1]*((nsnN/.n->3)-(nsnN/.n->4)+(nsnN/.
n->5)+(nsnN/.n->6)-

(nsnN/.n->7)+(nsnN/.n->8)));

sg5=(Exp[p*4*T/K1]*((nsnN/.n->4)-(nsnN/.n->5)-(nsnN/.
n->6)+(nsnN/.n->7)-

(nsnN/.n->8)+(nsnN/.n->9)));

sg6=-(Exp[p*5*T/K1]*((nsnN/.n->5)+(nsnN/.n->6)-(nsnN/.
n->7)+(nsnN/.n->8)-

(nsnN/.n->9)+(nsnN/.n->10)));

In this cell, **sg1** corresponds to the initial point $t_0 = 0$, **sg2** corresponds to the initial point $t_0 = t1$, **sg3** corresponds to the initial point $t_0 = 2t1$, and so on.

Now we define the Laplace transform of currents:

Iu:=E1*sg1/(1-Exp[-p*T/2])/(p*L+R);

Ic:=E1*Lsn2/(p*L+R)

In the first row we use the expression of the Laplace transform for a periodic function $f(t) = f(t+\frac{T}{2})$, which has the form

$$F(p) = \frac{\int_0^{T/2} f(t)e^{-pt}dt}{1-e^{-p\frac{T}{2}}}$$

for the period $\frac{T}{2}$.

According to the method described in Section 2.2, the steady-state process can be determined by calculating the expression

$$i_f(t) = \tilde{i}_n(t) + \tilde{i}_f(t) - i_n(t), \tag{2.4}$$

where $\tilde{i}_n(t)$ is the natural, and $\tilde{i}_f(t)$ the forced response determined for the continuous function $u_c(t) = (-1)^n E \sin(\omega(t+nt_1))$; and $i_n(t)$ is the natural response determined for the voltage $u(t)$.

In the next cell we form natural and forced responses for continuous and input functions:

α=R/L;

p1=I*ω;

i1:=Residue[Iu*Exp[p*t],{p,- α }];

i2:=Simplify[Factor[ExpToTrig[Residue[Ic*Exp[p*t],{p,p1}]+

Residue[Ic*Exp[p*t],{p,-p1}]]]]

i3:=Residue[Ic*Exp[p*t],{p,- α }]

In this cell, **i1** corresponds to $i_n(t)$, **i2** corresponds to $\tilde{i}_f(t)$, and **i3** corresponds to $\tilde{i}_n(t)$.

Let us plot the time diagram of the steady-state current $i(t)$. At first we enter the values of the parameters:

n=0;

K1=12;

E1=310.0;

R=20.0;

L=0.04;

T=20*10^(-3);

ω=2*Pi/T;

t1=T/K1;

Plotting of the current for the six intervals is made as follows:

On the first interval we use expressions of the currents defined in the previous cell:

p1i=Plot[-i1+i2+i3,{t,0,t1},AxesLabel->{"t","i"},DisplayFunction->Identity]

On the second interval,

n=1;

Iu=E1*sg2/(1-Exp[-p*T/2])/(p*L+R);

Ic=-(E1*Lsn2/(p*L+R));

isum=ReplaceAll[(-i1+i2+i3),t->t-n*T/K1];

**p2i=Plot[isum,{t, n*t1,(n+1)*t1},AxesLabel->{"t","i"},DisplayFunction
->Identity]**

we introduce the function **Iu** and **Ic** for the currents taking into account that the continuous function sin(ωt) is negative.

On the third interval,

n=2;

Iu:=E1*sg3/(1-Exp[-p*T/2])/(p*L+R);

Ic=(E1*Lsn2/(p*L+R));

isum=ReplaceAll[(-i1+i2+i3),t->t-n*T/K1];

p3i=Plot[isum,{t,n*t1,(n+1)*t1},AxesLabel->{"t","i"},DisplayFunction ->Identity]

On the fourth interval,

n=3;

Iu:=E1*sg4/(1-Exp[-p*T/2])/(p*L+R);

Ic=-(E1*Lsn2/(p*L+R));

isum=ReplaceAll[(-i1+i2+i3),t->t-n*T/K1];

p4i=Plot[isum,{t,n*t1,(n+1)*t1},AxesLabel->{"t","i"},DisplayFunction ->Identity]

On the fifth interval,

n=4;

Iu:=E1*sg5/(1-Exp[-p*T/2])/(p*L+R);

Ic=(E1*Lsn2/(p*L+R));

isum=ReplaceAll[(-i1+i2+i3),t->t-n*T/K1];

p5i=Plot[isum,{t,n*t1,(n+1)*t1},AxesLabel->{"t","i"},DisplayFunction ->Identity]

On the sixth interval,

n=5;

Iu:=E1*sg6/(1-Exp[-p*T/2])/(p*L+R);

Ic=-(E1*Lsn2/(p*L+R));

isum=ReplaceAll[(-i1+i2+i3),t->t-n*T/K1];

p6i=Plot[isum,{t,n*t1,(n+1)*t1},AxesLabel->{"t","i"},DisplayFunction ->Identity]

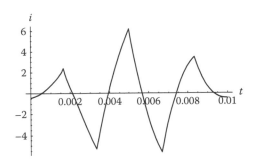

FIGURE 2.32
Time diagram of the steady-state current (*i* in amperes, time *t* in seconds).

We shift the beginning of the sum of currents **isum** at the point **t->t-n*T/K1**, corresponding to the beginning of a proper time interval.

Combining these graphs and outputting their results are realized using the function

Show[p1i,p2i,p3i,p4i,p5i,p6i,DisplayFunction->$DisplayFunction]

The time diagram of the steady-state current is presented in Figure 2.32.

Let us use Mathematica's tools to find the solution for the given problem. We define the voltage *u*(*t*) in the following way:

f1:=E1*((-1)^(Floor[2*t/(T)]))*((-1)^(Floor[K1*t/(T)]))*Sin[ω *t];

The function **Floor[.]** gives the greatest integer less than or equal to an argument. We use the expression **((-1)^(Floor[2*t/(T)]))** to rectify the sinusoidal signal, whereas we use the expression **((-1)^(Floor[K1*t/(T)]))** to multiply the rectified sinusoidal signal by the square wave sgn(sin($K\omega t$)). One can see that using the function **Plot[f1,{t,0,T}]** we obtain the same graph as in Figure 2.31.

Now we use Mathematica to solve Equation 2.45:

fnd = NDSolve[{i'[t] ==   R/L * i[t] + f1/L, i[0] == 0}, i,

{t, 0, 5 * T}, MaxSteps > 10000]

In this function, we choose the time interval equal to the five periods. Since the time constant $R/L = 2 \cdot 10^{-3}$ and period $T = 20 \cdot 10^{-3}$, this allows the obtaining of the steady-state process in the last of the intervals.

To compare the results for solving the differential Equation 2.45 by the considered and numerical methods, we can use the function

Plot[Evaluate[y[t]/.fnd],{t,4*T,5*T}]

It is not difficult to ensure that we obtain the same graph as in Figure 2.32.

2.8 Calculation of Processes in the Three-Phase Symmetric Matrix-Reactance Converter

Matrix-reactance converters have some properties that allow their efficient use in three-phase power supply systems. These properties are based on their capabilities in changing amplitudes and frequencies of output voltages and currents.

Let us consider a three-phase system with Buck-Boost and matrix converters (Korotyeyev and Fedyczak, 2008a) as shown in Figure 2.33. A modulation of the matrix converter switches is realized by pulse width modulation (PWM). The control strategy of the proposed matrix-reactance frequency converter (MRFC), in general form, is illustrated in Figure 2.34. In each sequence period T_S there are two time intervals, t_S and t_L. In the interval t_S, the synchronous-connected switches (SCS) are off, whereas the matrix-connected

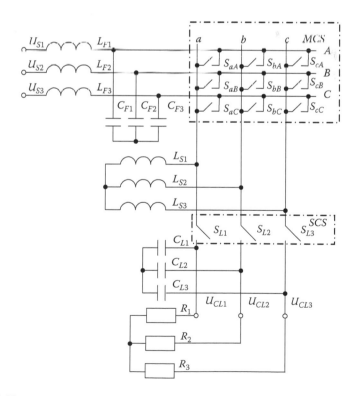

FIGURE 2.33
Matrix-reactance converter system, MCS—matrix-connected switches, SCS—synchronous-connected switches. (Data from Korotyeyev I. Ye. and Fedyczak Z., 2008b. With permission.)

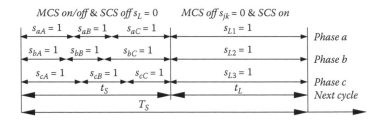

FIGURE 2.34
General form of the control strategy. (Data from Korotyeyev I. Ye. and Fedyczak Z., 2008b. With permission.)

switches (MCS) are switching in accordance with a control strategy. At those switching times, s_{jk} satisfy the condition

$$s_{j1} + s_{j2} + s_{j3} = 1 \quad \text{for} \quad j = 1, 2, 3.$$

In the interval t_L, the MCS are off, whereas the SCS are on. The MCS are controlled in line with the classical control strategy (Venturini and Alesina, 1980). For such a converter, MCS output voltages u_a, u_b, and u_c, and input currents i_A, i_B, i_C are formed as follows:

$$\begin{vmatrix} u_a \\ u_b \\ u_c \end{vmatrix} = M(t) \begin{vmatrix} u_A \\ u_B \\ u_C \end{vmatrix};$$

$$\begin{vmatrix} i_A \\ i_B \\ i_C \end{vmatrix} = M^T(t) \begin{vmatrix} i_a \\ i_b \\ i_c \end{vmatrix},$$

$$M(t) \begin{pmatrix} d_{aA} & d_{aB} & d_{aC} \\ d_{bA} & d_{bB} & d_{bC} \\ d_{cA} & d_{cB} & d_{cC} \end{pmatrix}, \qquad (2.46)$$

where
$d_{aA} = d_{bB} = d_{cC} = D_S(1 + 2q\cos(\omega_m t))/3;$
$d_{aB} = d_{cA} = d_{bC} = D_S(1 + 2q\cos(\omega_m t - 2\pi/3))/3;$
$d_{aC} = d_{bA} = d_{cB} = D_S(1 + 2q\cos(\omega_m t - 4\pi/3))/3;$
$\omega_m = \omega_L - \omega;$ ω_L is the pulsation of a voltage on a load; ω is a frequency of the supply voltage; T is the symbol of the transposition; q is the voltage gain; and $D_S = \frac{t_S}{T_S}$ is the duty ratio of the SCS.

Assuming that all switches are ideal, inductors and capacitors are linear, and, in order to simplify calculations, one uses the averaged operator (Korotyeyev and Fedyczak, 2002), and then the processes in such a system are described by the matrix differential equation

$$\frac{dX}{dt} = A(t)X + B(t), \tag{2.47}$$

where $X^T = (I_{LF1}\ I_{LF2}\ I_{LF3}\ I_{LS1}\ I_{LS2}\ I_{LS3}\ U_{CF1}\ U_{CF2}\ U_{CF3}\ U_{CL1}\ U_{CL2}\ U_{CL3})$; I_{LF1}, I_{LF2}, I_{LF3} are the currents in inductors L_{F1}, L_{F2}, L_{F3}; I_{LS1}, I_{LS2}, I_{LS3} are the currents in inductors L_{S1}, L_{S2}, L_{S3}; U_{CF1}, U_{CF2}, U_{CF3} are the voltages across capacitors C_{F1}, C_{F2}, C_{F3}; and U_{CL1}, U_{CL2}, U_{CL3} are the voltages across capacitors C_{L1}, C_{L2}, C_{L3}.

$A(t) =$

$$
\begin{pmatrix}
\frac{-R_{F1}}{L_{F1}} & 0 & 0 & 0 & 0 & 0 & \frac{-1}{L_{F1}} & 0 & 0 & 0 & 0 & 0 \\
0 & \frac{-R_{F2}}{L_{F2}} & 0 & 0 & 0 & 0 & 0 & \frac{-1}{L_{F2}} & 0 & 0 & 0 & 0 \\
0 & 0 & \frac{-R_{F3}}{L_{F3}} & 0 & 0 & 0 & 0 & 0 & \frac{-1}{L_{F3}} & 0 & 0 & 0 \\
0 & 0 & 0 & \frac{-R_{S1}}{L_{S1}} & 0 & 0 & \frac{d_{aA}}{L_{S1}} & \frac{d_{aB}}{L_{S1}} & \frac{d_{aC}}{L_{S1}} & \frac{1-D_S}{L_{S1}} & 0 & 0 \\
0 & 0 & 0 & 0 & \frac{-R_{S2}}{L_{S2}} & 0 & \frac{d_{bA}}{L_{S2}} & \frac{d_{bB}}{L_{S2}} & \frac{d_{bC}}{L_{S2}} & 0 & \frac{1-D_S}{L_{S2}} & 0 \\
0 & 0 & 0 & 0 & 0 & \frac{-R_{S3}}{L_{S3}} & \frac{d_{cA}}{L_{S3}} & \frac{d_{cB}}{L_{S3}} & \frac{d_{cC}}{L_{S3}} & 0 & 0 & \frac{1-D_S}{L_{S3}} \\
\frac{1}{C_{F1}} & 0 & 0 & \frac{-d_{aA}}{C_{F1}} & \frac{-d_{bA}}{C_{F1}} & \frac{-d_{cA}}{C_{F1}} & 0 & 0 & 0 & 0 & 0 & 0 \\
0 & \frac{1}{C_{F2}} & 0 & \frac{-d_{aB}}{C_{F2}} & \frac{-d_{bB}}{C_{F2}} & \frac{-d_{cB}}{C_{F2}} & 0 & 0 & 0 & 0 & 0 & 0 \\
0 & 0 & \frac{1}{C_{F3}} & \frac{-d_{aC}}{C_{F3}} & \frac{-d_{bC}}{C_{F3}} & \frac{-d_{cC}}{C_{F3}} & 0 & 0 & 0 & 0 & 0 & 0 \\
0 & 0 & 0 & \frac{D_S-1}{C_{S1}} & 0 & 0 & 0 & 0 & 0 & \frac{-1}{R_1 C_{S1}} & 0 & 0 \\
0 & 0 & 0 & 0 & \frac{D_S-1}{C_{S2}} & 0 & 0 & 0 & 0 & 0 & \frac{-1}{R_2 C_{S2}} & 0 \\
0 & 0 & 0 & 0 & 0 & \frac{D_S-1}{C_{S3}} & 0 & 0 & 0 & 0 & 0 & \frac{-1}{R_3 C_{S3}}
\end{pmatrix}
$$

$$B^T(t) = \left(\frac{U_1}{L_{F1}}\cos(\omega t)\ \frac{U_2}{L_{F2}}\cos(\omega t + 2\pi/3)\ \frac{U_3}{L_{F3}}\cos(\omega t + 4\pi/3)\ 0\ 0\ 0\ 0\ 0\ 0\ 0\ 0\ 0 \right).$$

U_1, U_2, U_3 are the amplitudes of supply voltages.

We assume that the system is symmetrical, that is,

$$R_{F1} = R_{F2} = R_{F3} = R_F; \quad R_{S1} = R_{S2} = R_{S3} = R_S;$$

$$L_{F1} = L_{F2} = L_{F3} = L_F; \quad L_{S1} = L_{S2} = L_{S3} = L_S;$$

$$C_{F1} = C_{F2} = C_{F3} = C_F; \quad C_{L1} = C_{L2} = C_{L3} = C_L;$$

$$R_1 = R_2 = R_3 = R \quad \text{and} \quad U_1 = U_2 = U_3 = U;$$

where R_F, and R_S are the resistances of inductors L_F and L_S; and R is the load resistance.

In consequence to the modulation strategy, the processes in converter systems are described by nonstationary differential equations. Calculations of transient and steady-state processes in such systems can be realized by numerical means. Based on the assumption of symmetry, steady-state and transient processes can be found analytically.

2.8.1 Double-Frequency Complex Function Method

Let us find steady-state processes in the matrix-reactance converter. Since in the matrix $A(t)$ and in the vector $B(t)$ there are signals that depend on two frequencies, we introduce a double-frequency complex function model and describe the state variable vector $\mathbf{x}(t)$ as follows:

$$\underline{X}^T = \left(\underline{I}_{LF1} e^{j\omega t} \ \underline{I}_{LF2} e^{j\omega t} \ \underline{I}_{LF3} e^{j\omega t} \ \underline{I}_{LS1} e^{j\omega_L t} \ \underline{I}_{LS2} e^{j\omega_L t} \ \underline{I}_{LS3} e^{j\omega_L t} \right.$$

$$\left. \underline{U}_{CF1} e^{j\omega t} \ \underline{U}_{CF2} e^{j\omega t} \ \underline{U}_{CF3} e^{j\omega t} \ \underline{U}_{CL1} e^{j\omega_L t} \ \underline{U}_{CL2} e^{j\omega_L t} \ \underline{U}_{CL3} e^{j\omega_L t} \right.$$

where $\underline{I}_{LF1}$, $\underline{I}_{LF2}$, $\underline{I}_{LF3}$ are amplitudes of currents in inductors L_{F1}, L_{F2}, L_{F3}; $\underline{I}_{LS1}$, $\underline{I}_{LS2}$, $\underline{I}_{LS3}$ are the amplitudes of currents in inductors L_{S1}, L_{S2}, L_{S3}; $\underline{U}_{CF1}$, $\underline{U}_{CF2}$, $\underline{U}_{CF3}$ are the amplitudes of voltages across capacitors C_{F1}, C_{F2}, C_{F3}; and $\underline{U}_{CL1}$, $\underline{U}_{CL2}$, $\underline{U}_{CL3}$ are amplitudes of voltages across capacitors C_{L1}, C_{L2}, C_{L3}.

For symmetry of both the MRFC circuit and the supply source, the state variables can be described as follows:

$$\underline{I}_{LF2} = \underline{I}_{LF1} e^{j2\pi/3}; \quad \underline{I}_{LF3} = \underline{I}_{LF1} e^{j4\pi/3}; \quad \underline{I}_{LS2} = \underline{I}_{LS1} e^{j2\pi/3}; \quad \underline{I}_{LS3} = \underline{I}_{LS1} e^{j4\pi/3};$$

$$\underline{U}_{CF2} = \underline{U}_{CF1} e^{j2\pi/3}; \quad \underline{U}_{CF3} = \underline{U}_{CF1} e^{j4\pi/3}; \quad \underline{U}_{CL2} = \underline{U}_{CL1} e^{j2\pi/3}; \quad \underline{U}_{CL3} = \underline{U}_{CL1} e^{j4\pi/3}.$$

Then, the vectors $X(t)$ and $B(t)$ can be described as

$$\underline{X}^T = \left(\underline{I}_{LF1} e^{j\omega t} \ \underline{I}_{LF1} e^{j(\omega t + 2\pi/3)} \ \underline{I}_{LF1} e^{j(\omega t + 4\pi/3)} \ \underline{I}_{LS1} e^{j\omega_L t} \ \underline{I}_{LS1} e^{j(\omega_L t + 2\pi/3)} \ \underline{I}_{LS1} e^{j(\omega_L t + 4\pi/3)} \right.$$

$$\left. \underline{U}_{CF1} e^{j\omega t} \ \underline{U}_{CF1} e^{j(\omega t + 2\pi/3)} \ \underline{U}_{CF1} e^{j(\omega t + 4\pi/3)} \ \underline{U}_{CL1} e^{j\omega_L t} \ \underline{U}_{CL1} e^{j(\omega_L t + 2\pi/3)} \ \underline{U}_{CL1} e^{j(\omega_L t + 4\pi/3)} \right);$$

$$B^T(t) = \left(\frac{U}{L_F} e^{j\omega t} \ \frac{U}{L_F} e^{j(\omega t + 2\pi/3)} \ \frac{U}{L_F} e^{j(\omega t + 4\pi/3)} \ 0 \ 0 \ 0 \ 0 \ 0 \ 0 \ 0 \ 0 \ 0 \right). \quad (2.48)$$

Taking the derivative of $X(t)$, we get

$$\frac{d\underline{X}^T}{dt} = \left(\omega \underline{I}_{LF1} e^{j\omega t} \ \omega \underline{I}_{LF1} e^{j(\omega t + 2\pi/3)} \ \omega \underline{I}_{LF1} e^{j(\omega t + 4\pi/3)} \ \omega_L \underline{I}_{LS1} e^{j\omega_L t} \right.$$

$$\omega_L \underline{I}_{LS1} e^{j(\omega_L t + 2\pi/3)} \ \omega_L \underline{I}_{LS1} e^{j(\omega_L t + 4\pi/3)} \ \omega \underline{U}_{CF1} e^{j\omega t} \ \omega \underline{U}_{CF1} e^{j(\omega t + 2\pi/3)} \quad (2.49)$$

$$\left. \omega \underline{U}_{CF1} e^{j(\omega t + 4\pi/3)} \ \omega_L \underline{U}_{CL1} e^{j\omega_L t} \ \omega_L \underline{U}_{CL1} e^{j(\omega_L t + 2\pi/3)} \ \omega_L \underline{U}_{CL1} e^{j(\omega_L t + 4\pi/3)} \right).$$

Substituting (2.48) and (2.49) into (2.47), and multiplying the matrix $A(t)$ by the vector $X(t)$, we obtain 12 linear equations in which each row has the same factor ($e^{j\omega t}$, $e^{j(\omega t + 2\pi/3)}$, $e^{j(\omega t + 4\pi/3)}$ or $e^{j\omega_L t}$, $e^{j(\omega_L t + 2\pi/3)}$, $e^{j(\omega_L t + 4\pi/3)}$). It turns out that, after cancellation of common factors, we can choose only four independent equations. In matrix form these equations can be written as follows:

$$\underline{X}' = \underline{A}\underline{X} + \underline{B}, \quad (2.50)$$

where

$$\underline{X}' = j \begin{pmatrix} \omega \underline{I}_{LF1} \\ \omega_L \underline{I}_{LS1} \\ \omega \underline{U}_{CF1} \\ \omega_L \underline{U}_{CL1} \end{pmatrix}, \quad \underline{X} = \begin{pmatrix} \underline{I}_{LF1} \\ \underline{I}_{LS1} \\ \underline{U}_{CF1} \\ \underline{U}_{CL1} \end{pmatrix}, \quad \underline{B} = \begin{pmatrix} \dfrac{U}{L_F} \\ 0 \\ 0 \\ 0 \end{pmatrix},$$

$$\underline{A} = \begin{pmatrix} \dfrac{-R_F}{L_F} & 0 & \dfrac{-1}{L_F} & 0 \\ 0 & \dfrac{-R_S}{L_S} & \dfrac{D_S q}{L_S} & \dfrac{1 - D_S}{L_S} \\ \dfrac{1}{C_F} & \dfrac{-D_S q}{C_F} & 0 & 0 \\ 0 & \dfrac{D_S - 1}{C_L} & 0 & \dfrac{-1}{R C_L} \end{pmatrix}.$$

Solving (2.50) for $\underline{X}$ yields

$$\underline{X} = (jI_\omega - \underline{A})^{-1}\underline{B},$$ (2.51)

where

$$I_\omega = \begin{pmatrix} \omega & 0 & 0 & 0 \\ 0 & \omega_L & 0 & 0 \\ 0 & 0 & \omega & 0 \\ 0 & 0 & 0 & \omega_L \end{pmatrix}.$$

The solution to (2.51) gives the components of the vector $\underline{X}$ expressed by

$$\underline{I}_{LF1} = \frac{jU\{C_F R\omega(1-2D_S) + C_F\omega(R_S + jL_S\omega_L)(1 + jC_L R\omega_L) + D_S^2[C_F R\omega + q^2(-j + C_L R\omega_L)]\}}{\Delta};$$

$$\underline{I}_{CL1} = \frac{UqD_S(1+jC_L R\omega_L)}{\Delta}, \quad \underline{U}_{CL1} = \frac{UqRD_S(D_S-1)}{\Delta},$$ (2.52)

$$\underline{U}_{CF1} = \frac{U[R_S + jL_S\omega_L + R[1 - 2D_S + D_S^2 + C_L\omega_L(jR_S - L_S\omega_L)]]}{\Delta},$$

where

$$\Delta = D_S^2 q^2(R_F + jL_F\omega) + [1 + C_F\omega(jR_F - L_F\omega)](R_S + jL_S\omega_L) +$$

$$R\{2D_S[-1 + C_F\omega(-jR_F + L_F\omega)] + D_S^2[1 + C_F\omega(jR_F - L_F\omega) + jC_L q^2 \cdot$$

$$(R_F + jL_F\omega)\omega_L] + [-1 + C_F\omega(-jR_F + L_F\omega)][-1 + C_L\omega_L(-jR_S + L_S\omega_L)]\}.$$

Instantaneous values of currents and voltages are obtained by multiplication of complex variables $\underline{I}_{LF1}$, $\underline{I}_{LC1}$, $\underline{U}_{CF1}$, and $\underline{U}_{CL1}$ by the function of either $e^{j\omega t}$ or $e^{j\omega_L t}$:

$$I_{LF1}(t) = \text{Re}\left(\underline{I}_{LF1}e^{j\omega t}\right); \quad I_{CL1}(t) = \text{Re}\left(\underline{I}_{LC1}e^{j\omega_L t}\right);$$

$$U_{CF1}(t) = \text{Re}\left(\underline{U}_{CF1}e^{j\omega t}\right); \quad U_{CL1}(t) = \text{Re}\left(\underline{U}_{CL1}e^{j\omega_L t}\right).$$ (2.53)

Expressions (2.53) are a solution of the set (2.47) for the steady-state process. Let us use Mathematica for solving the equation set (2.47). In the cell

ωm=ωL-ω;

daA:=Ds*(1+2*qu*Cos[ωm*t])/3;

daB:=Ds*(1+2*qu*Cos[ωm*t-2*Pi/3])/3;

$$daC:=Ds*(1+2*qu*Cos[\omega m*t-4*Pi/3])/3;$$
$$dbA:=Ds*(1+2*qu*Cos[\omega m*t-4*Pi/3])/3;$$
$$dbB:=Ds*(1+2*qu*Cos[\omega m*t])/3;$$
$$dbC:=Ds*(1+2*qu*Cos[\omega m*t-2*Pi/3])/3;$$
$$dcA:=Ds*(1+2*qu*Cos[\omega m*t-2*Pi/3])/3;$$
$$dcB:=Ds*(1+2*qu*Cos[\omega m*t-4*Pi/3])/3;$$
$$dcC:=Ds*(1+2*qu*Cos[\omega m*t])/3;$$

we define the components of the matrix $M(t)$ as in (2.46). Next, we define the vectors $X(t)$, $B(t)$, and the matrix $A(t)$ as follows:

$$Xin = \begin{pmatrix} Is*Exp[I*(\ *t)] \\ Is*Exp[I*(\ *t+2*Pi/3)] \\ Is*Exp[I*(\ *t+4*Pi/3)] \\ Ic*Exp[I*(\ L*t)] \\ Ic*Exp[I*(\ L*t+4*Pi/3)] \\ Ic*Exp[I*(\ L*t+4*Pi/3)] \\ Uc*Exp[I*(\ *t)] \\ Uc*Exp[I*(\ *t+2*Pi/3)] \\ Uc*Exp[I*(\ *t+4*Pi/3)] \\ Ul*Exp[I*(\ L*t)] \\ Ul*Exp[I*(\ L*t+2*Pi/3)] \\ Ul*Exp[I*(\ L*t+4*Pi/3)] \end{pmatrix};$$

$$B(t) = \begin{pmatrix} U*Exp[I*(\ *t)]/Lf \\ U*Exp[I*(\ *t+2*Pi/3)]/Lf \\ U*Exp[I*(\ *t+4*Pi/3)]/Lf \\ 0 \\ 0 \\ 0 \\ 0 \\ 0 \\ 0 \\ 0 \\ 0 \\ 0 \end{pmatrix};$$

A(t) =

$$
\begin{pmatrix}
\dfrac{-Rf}{Lf} & 0 & 0 & 0 & 0 & 0 & \dfrac{-1}{Lf} & 0 & 0 & 0 & 0 & 0 \\
0 & \dfrac{-Rf}{Lf} & 0 & 0 & 0 & 0 & 0 & \dfrac{-1}{Lf} & 0 & 0 & 0 & 0 \\
0 & 0 & \dfrac{-Rf}{Lf} & 0 & 0 & 0 & 0 & 0 & \dfrac{-1}{Lf} & 0 & 0 & 0 \\
0 & 0 & 0 & \dfrac{-Rs}{Ls} & 0 & 0 & \dfrac{d_{aA}}{Ls} & \dfrac{d_{aB}}{Ls} & \dfrac{d_{aC}}{Ls} & \dfrac{1-D_S}{Ls} & 0 & 0 \\
0 & 0 & 0 & 0 & \dfrac{-Rs}{Ls} & 0 & \dfrac{d_{bA}}{Ls} & \dfrac{d_{bB}}{Ls} & \dfrac{d_{bC}}{Ls} & 0 & \dfrac{1-D_S}{Ls} & 0 \\
0 & 0 & 0 & 0 & 0 & \dfrac{-Rs}{Ls} & \dfrac{d_{cA}}{Ls} & \dfrac{d_{cB}}{Ls} & \dfrac{d_{cC}}{Ls} & 0 & 0 & \dfrac{1-D_S}{Ls} \\
\dfrac{1}{Cf} & 0 & 0 & \dfrac{-d_{aA}}{Cf} & \dfrac{-d_{bA}}{Cf} & \dfrac{-d_{cA}}{Cf} & 0 & 0 & 0 & 0 & 0 & 0 \\
0 & \dfrac{1}{Cf} & 0 & \dfrac{-d_{aB}}{Cf} & \dfrac{-d_{bB}}{Cf} & \dfrac{-d_{cB}}{Cf} & 0 & 0 & 0 & 0 & 0 & 0 \\
0 & 0 & \dfrac{1}{Cf} & \dfrac{-d_{aC}}{Cf} & \dfrac{-d_{bC}}{Cf} & \dfrac{-d_{cC}}{Cf} & 0 & 0 & 0 & 0 & 0 & 0 \\
0 & 0 & 0 & \dfrac{D_S-1}{Cl} & 0 & 0 & 0 & 0 & 0 & \dfrac{-1}{RCl} & 0 & 0 \\
0 & 0 & 0 & 0 & \dfrac{D_S-1}{Cl} & 0 & 0 & 0 & 0 & 0 & \dfrac{-1}{RCl} & 0 \\
0 & 0 & 0 & 0 & 0 & \dfrac{D_S-1}{Cl} & 0 & 0 & 0 & 0 & 0 & \dfrac{-1}{RCl}
\end{pmatrix} ;
$$

$$\tag{2.54}$$

The symbols **Xin, Is, Ic, Uc, Ul** correspond to $\underline{X}$, I_{LF1}, I_{CL1}, U_{CF1}, U_{CL1}, respectively.

In the row

$$\text{DXin} = \partial_t \text{Xin};$$

the derivative of $\underline{X}$ is calculated. In order to show that some parts of

$$\frac{d\underline{X}}{dt} - A(t)\underline{X} - B(t) = 0 \tag{2.55}$$

are the same, we calculate the left part of this equation:

XE=Simplify[TrigToExp[A11.Xin+E1]-DXin];

and cancel each row by $e^{-j\omega t}$ or $e^{-j\omega_L t}$ factors:

For[n=1,n<=12,If[n<=3 || n>=7&&n<=9,

AXL[n]=Simplify[Part[Part[XE*Exp[-I*ω*t],n],1]],

AXL[n]=Simplify[Part[Part[XE*Exp[-I* ωL*t],n],1]]];n++];

Now we will show that some parts of (2.55) are the same. We output the first three parts of the obtained equations:

$$\textbf{For[n=1,n<=3,Print[AXL[n]];n++];}$$

$$-\frac{IsRf - U + Uc + iIsLf\omega}{Lf}$$

$$-\frac{(-1)^{2/3}(-U + Uc + Is(Rf + iLf\omega))}{Lf}$$

$$-\frac{(-1)^{1/3}(-U + Uc + Is(Rf + iLf\omega))}{Lf}$$

After equating these expressions to zero, the constant factors are canceled, and we obtain three identical equations.

In the next cell we find a solution to (2.50). At first we define the matrix I_ω and vector $\underline{B}$ (denoted by **B4**):

$$\textbf{I}\omega\textbf{=IdentityMatrix[4];}$$

$$\textbf{I}\omega\textbf{[[1,1]]=I*}\omega\textbf{;}$$

$$\textbf{I}\omega\textbf{[[2,2]]=I*}\omega\textbf{L;}$$

$$\textbf{I}\omega\textbf{[[3,3]]=I*}\omega\textbf{;}$$

$$\textbf{I}\omega\textbf{[[4,4]]=I*}\omega\textbf{L;}$$

$$\textbf{B4} = \begin{pmatrix} U/Lf \\ 0 \\ 0 \\ 0 \end{pmatrix};$$

Elements of the matrix $\underline{A}$ denoted by **A4** are determined by the calculation $A(t)\underline{X}(t)$

$$\textbf{AX=Simplify[TrigToExp[A11.Xin]];}$$

cancellation by $e^{-j\omega t}$ or $e^{-j\omega_L t}$ factors in rows

$$\textbf{For[n=1,n<=12,If[n<=3 || n>=7\&\&n<=9,}$$

$$\textbf{XL[n]=Simplify[Part[Part[AX*Exp[-I*}\omega\textbf{*t],n],1]],}$$

$$\textbf{XL[n]=Simplify[Part[Part[AX*Exp[-I*}\ \omega\textbf{L*t],n],1]]];n++];}$$

and extraction of coefficients from rows 1, 4, 7, 10:

$$ZS=\{Is,Ic,Uc,Ul\};$$

$$A4=IdentityMatrix[4];$$

$$For[n=1,n<=4,$$

$$For[m=1,m<=4,A4[[n,m]]=Coefficient[XL[3*(n-1)+1],ZS[[m]]]];m++];n++];$$

According to (2.51), we find the solution

$$solIA=Simplify[Inverse[I\omega-A4].B4];$$

By transforming and simplifying the components in the expression **solIA**, we obtain solutions in the form (2.52).

Using Mathematica we can find the same solution in the following way:

$$usl=Table[Part[XE[[n]],1]==0,\{n,1,12\}];$$

$$solX=Simplify[Solve[usl,\{Is,Ic,Uc,Ul\}]];$$

In the first row we generate the set of equations, and in the second row we find the solution.

Let us find the steady-state values of currents and voltages. In the cell we enter the parameters of the circuit elements, the supply voltage, and control signals:

$$Rf=0.01;$$

$$Rs=0.01;$$

$$Lf=0.0005;$$

$$Ls=0.0005;$$

$$Cf=50*10^{(-6)};$$

$$Cl=50*10^{(-6)};$$

$$R=10.0;$$

$$U=230.0;$$

$$T=1/50.0;$$

$$\omega=2*Pi/T;$$

$$TL=1/25.0;$$

$$\omega L:=2*Pi/TL;$$

$$qu=0.5;$$

$$Ds=0.7;$$

Then we select the components that correspond to currents I_{LF1}, and I_{LS1} and voltages U_{CF1}, and U_{CL1}:

$$Is1:=Part[solIA[[1]],1];$$

$$Ic1:=Part[solIA[[2]],1];$$

$$Uc1:=Part[solIA[[3]],1];$$

$$Ul1:=Part[solIA[[4]],1];$$

The solutions (2.53) are defined as follows:

$$p1=Re[Is1*Exp[I*\omega*t]];$$

$$p2=Re[Ic1*Exp[I*\omega L*t]];$$

$$p3=Re[Uc1*Exp[I*\omega*t]];$$

$$p4=Re[Ul1*Exp[I*\omega L*t]];$$

We also define a function of the first phase supply voltage:

$$psupp:=U*Cos[\omega *t];$$

Time diagrams of the steady-state processes for currents and voltages are outputted with the help of the functions

$$Plot[\{p1,p2\},\{t,0,2*T\}];$$

$$Plot[\{p3,p4,psupp\},\{t,0,2*T\}];$$

These time diagrams are shown in Figures 2.35 and 2.36.

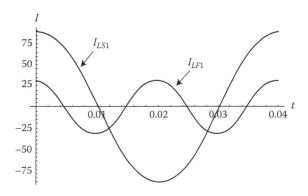

FIGURE 2.35

The steady-state currents I_{LF1} and I_{LS1} in inductors (I_{LF1} and I_{LS1} in amperes, time t in seconds).

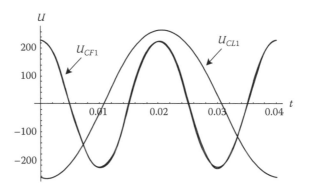

FIGURE 2.36
The steady-state voltages across capacitors U_{CL1}, U_{CF1} and supply voltage U_{S1} (U_{CL1}, U_{CF1}, and U_{S1} in volts, time t in seconds).

The graphs of the relative magnitude and phase shift of the load voltage as functions of D_S and q obtained by

$$\textbf{Plot3D[Abs[Ul1]/U,\{Ds,0.1,0.99\},\{qu,0.1,0.5\},}$$

$$\textbf{AxesLabel->\{"Ds","q","UcL1/U"\},Shading->False];}$$

$$\textbf{Plot3D[Arg[Ul1],\{Ds,0.1,0.7\},\{qu,0.1,0.5\},}$$

$$\textbf{AxesLabel->\{"Ds","φUcL1"\},Shading->False];}$$

are presented correspondingly in Figures 2.37 and 2.38.

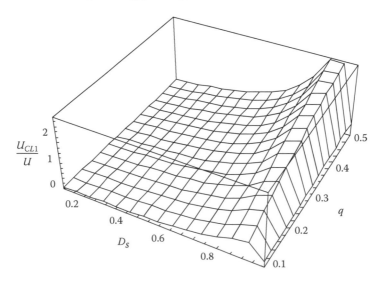

FIGURE 2.37
The relative magnitude of the load voltage versus parameters D_s and q.

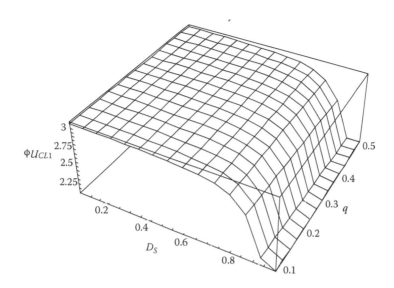

FIGURE 2.38
The phase shift of the load voltage versus parameters D_s and q.

From these figures one sees that regulation of the output voltage can be made by changing the parameters q and D_s while the phase shift can be made mainly by changing the parameter D_s.

The graphs of the relative magnitudes of the load voltage with respect to the supply voltage, and the input power factor versus the parameter D_s for different values of a period T_L obtained by

$$qu=0.5;$$

$$p025V=Plot[Abs[Ul1]/U,\{Ds,0.0,0.95\},AxesLabel->\{"D","UcL1/U"\},$$

$$DisplayFunction->Identity];$$

$$p025A=Plot[Cos[Arg[Is1]],\{Ds,0.0,0.95\},AxesLabel->\{"D","\lambda p"\},$$

$$DisplayFunction->Identity];$$

$$TL=1/75.0;$$

$$p075V=Plot[Abs[Ul1/U],\{Ds,0.0,0.95\},AxesLabel->\{"D","UcL1/U"\},$$

$$DisplayFunction->Identity];$$

$$p075A=Plot[Cos[Arg[Is1]],\{Ds,0.0,0.95\},AxesLabel->\{"D","\lambda p"\},$$

$$DisplayFunction->Identity];$$

$$TL=1/50.0;$$

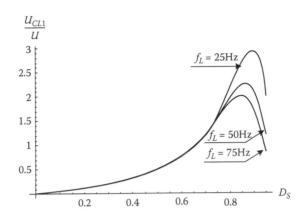

FIGURE 2.39
The relative magnitude of the load voltage versus parameter D_s for different T_L values.

p050V=Plot[Abs[Ul1]/U,{Ds,0.0,0.95},AxesLabel->{"D","UcL1/U"},

DisplayFunction->Identity];

p050A=Plot[Cos[Arg[Is1]],{Ds,0.0,0.95},AxesLabel->{"D","λp"},

DisplayFunction->Identity];

Show[{p025V,p050V,p075V},DisplayFunction->$DisplayFunction];

are presented in Figures 2.39 and 2.40. The power factor is defined as $\lambda_p = \cos(\phi_s)$, where ϕ_s is the phase shift between the voltage U_{S1} and the current I_{LF1}.

Comparing Figures 2.39 and 2.40, one can change the gain and have the power factor near to unity.

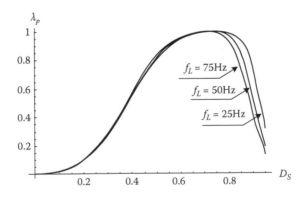

FIGURE 2.40
The input power factor versus parameter D_s for different T_L values.

2.8.2 Double-Frequency Transform Matrix Method

Let us find transient process in the symmetric matrix-reactance converter. We transform (2.47) with the use of the matrix (Korotyeyev and Fedyczak, 2008b)

$$K = \begin{pmatrix} K_S & 0 & 0 & 0 \\ 0 & K_L & 0 & 0 \\ 0 & 0 & K_S & 0 \\ 0 & 0 & 0 & K_L \end{pmatrix},$$

where

$$K_S = \begin{pmatrix} 1+\cos(\omega t + \varphi) & 1+\cos\left(\omega t + \dfrac{2\pi}{3} + \varphi\right) & 1+\cos\left(\omega t - \dfrac{2\pi}{3} + \varphi\right) \\ 1+\cos\left(\omega t + \dfrac{2\pi}{3} + \varphi\right) & 1+\cos\left(\omega t - \dfrac{2\pi}{3} + \varphi\right) & 1+\cos(\omega t + \varphi) \\ 1+\cos\left(\omega t - \dfrac{2\pi}{3} + \varphi\right) & 1+\cos(\omega t + \varphi) & 1+\cos\left(\omega t + \dfrac{2\pi}{3} + \varphi\right) \end{pmatrix};$$

$$K_L = \begin{pmatrix} 1+\cos(\omega_L t + \varphi) & 1+\cos\left(\omega_L t + \dfrac{2\pi}{3} + \varphi\right) & 1+\cos\left(\omega_L t - \dfrac{2\pi}{3} + \varphi\right) \\ 1+\cos\left(\omega_L t + \dfrac{2\pi}{3} + \varphi\right) & 1+\cos\left(\omega_L t - \dfrac{2\pi}{3} + \varphi\right) & 1+\cos(\omega_L t + \varphi) \\ 1+\cos\left(\omega_L t - \dfrac{2\pi}{3} + \varphi\right) & 1+\cos(\omega_L t + \varphi) & 1+\cos\left(\omega_L t + \dfrac{2\pi}{3} + \varphi\right) \end{pmatrix},$$

and φ is the phase shift.

Substituting X for KY yields

$$\frac{dK}{dt}Y + K\frac{dY}{dt} = A(t)KY + B(t),$$

where Y is the vector of transformed system variables. The matrix K is not singular. An inverse matrix

$$K_S^{-1} = \frac{1}{3\sqrt{2}} \cdot$$

$$\{\{-1-4\cos(\omega t + \varphi), -1+2\cos(\omega t + \varphi)+2\sqrt{3}\sin(\omega t + \varphi), -1+2\cos(\omega t + \varphi)$$
$$-2\sqrt{3}\sin(\omega t + \varphi)\},$$

$$\{-1+2\cos(\omega t + \varphi)+2\sqrt{3}\sin(\omega t + \varphi), -1+2\cos(\omega t + \varphi)-2\sqrt{3}\sin(\omega t + \varphi),$$
$$-1-4\cos(\omega t + \varphi),\}$$

$$\{-1+2\cos(\omega t + \varphi)-2\sqrt{3}\sin(\omega t + \varphi), -1-4\cos(\omega t + \varphi), -1+2\cos(\omega t + \varphi)$$
$$+2\sqrt{3}\sin(\omega t + \varphi)\}\}$$

An inverse matrix K_L^{-1} has a similar form.

Premultiplying this equation by the inverse matrix K^{-1} and taking into account that

$$\Omega = K^{-1}\frac{dK}{dt} = \begin{pmatrix} \Omega_S & 0 & 0 & 0 \\ 0 & \Omega_L & 0 & 0 \\ 0 & 0 & \Omega_S & 0 \\ 0 & 0 & 0 & \Omega_L \end{pmatrix}$$

we obtain

$$\frac{dY}{dt} = (K^{-1}A(t)K - \Omega)Y + K^{-1}B(t), \qquad (2.56)$$

where

$$\Omega_S = \begin{pmatrix} 0 & -\dfrac{\omega}{\sqrt{3}} & \dfrac{\omega}{\sqrt{3}} \\[2ex] \dfrac{\omega}{\sqrt{3}} & 0 & -\dfrac{\omega}{\sqrt{3}} \\[2ex] -\dfrac{\omega}{\sqrt{3}} & \dfrac{\omega}{\sqrt{3}} & 0 \end{pmatrix},$$

$$\Omega_L = \begin{pmatrix} 0 & -\dfrac{\omega_L}{\sqrt{3}} & \dfrac{\omega_L}{\sqrt{3}} \\[2ex] \dfrac{\omega_L}{\sqrt{3}} & 0 & -\dfrac{\omega_L}{\sqrt{3}} \\[2ex] -\dfrac{\omega_L}{\sqrt{3}} & \dfrac{\omega_L}{\sqrt{3}} & 0 \end{pmatrix}.$$

In (2.56) the matrix $K^{-1}A(t)K$ and the vector $K^{-1}B(t)$ do not depend on time. Denoting

$$K^{-1}A(t)K = A \quad \text{and} \quad K^{-1}B(t) = B,$$

Equation 2.56 can be rewritten as follows:

$$\frac{dY}{dt} = (A - \Omega)Y + B. \qquad (2.57)$$

By means of the transformation, the nonstationary matrix differential equation (2.47) describing processes in the three-phase matrix-reactance converter system has been transformed into the stationary differential equation (2.57). From this equation the solution is obtained in an ordinary way. In Equation 2.57 the matrix A has the forms

$$A =$$

$$
\begin{pmatrix}
\frac{-R_F}{L_F} & 0 & 0 & 0 & 0 & 0 & \frac{-1}{L_F} & 0 & 0 & 0 & 0 & 0 \\
0 & \frac{-R_F}{L_F} & 0 & 0 & 0 & 0 & 0 & \frac{-1}{L_F} & 0 & 0 & 0 & 0 \\
0 & 0 & \frac{-R_F}{L_F} & 0 & 0 & 0 & 0 & 0 & \frac{-1}{L_F} & 0 & 0 & 0 \\
0 & 0 & 0 & \frac{-R_S}{L_S} & 0 & 0 & \frac{a_1}{L_S} & \frac{a_2}{L_S} & \frac{u_3}{L_S} & \frac{1-D_S}{l_S} & 0 & 0 \\
0 & 0 & 0 & 0 & \frac{-R_S}{L_S} & 0 & \frac{a_3}{L_S} & \frac{a_1}{L_S} & \frac{a_2}{L_S} & 0 & \frac{1-D_S}{L_S} & 0 \\
0 & 0 & 0 & 0 & 0 & \frac{-R_S}{L_S} & \frac{a_2}{L_S} & \frac{a_3}{L_S} & \frac{a_1}{L_S} & 0 & 0 & \frac{1-D_S}{L_S} \\
\frac{1}{C_F} & 0 & 0 & \frac{-a_1}{C_F} & \frac{-a_3}{C_F} & \frac{-a_2}{C_F} & 0 & 0 & 0 & 0 & 0 & 0 \\
0 & \frac{1}{C_\Gamma} & 0 & \frac{-a_2}{C_F} & \frac{a_1}{C_F} & \frac{-u_3}{C_F} & 0 & 0 & U & 0 & 0 & U \\
0 & 0 & \frac{1}{C_F} & \frac{-a_3}{C_F} & \frac{-a_2}{C_F} & \frac{-a_1}{C_F} & 0 & U & 0 & 0 & 0 & 0 \\
0 & 0 & 0 & \frac{D_S-1}{C_S} & 0 & 0 & 0 & 0 & 0 & \frac{-1}{RC_S} & 0 & 0 \\
0 & 0 & 0 & 0 & \frac{D_S-1}{C_S} & 0 & 0 & 0 & 0 & 0 & \frac{-1}{RC_S} & 0 \\
0 & 0 & 0 & 0 & 0 & \frac{D_S-1}{C_S} & 0 & 0 & 0 & 0 & 0 & \frac{-1}{RC_S}
\end{pmatrix}
$$

where $a_1 = (1+2q)/3$; $a_2 = (1-q)/3$; $a_3 = (1-q)/3$.

Let us consider a more general case when the vector

$$
B^T(t) = \left(\frac{U}{L_F}\cos(\omega t + \psi) \quad \frac{U}{L_F}\cos(\omega t + 2\pi/3 + \psi) \quad \frac{U}{L_F}\cos(\omega t + 4\pi/3 + \psi)\,000000000 \right),
$$

where ψ is the phase shift of the supply voltages. Then vector

$$
B^T = \left(-\sqrt{2}\frac{U}{L_F}\cos(\psi - \varphi)\frac{U(\cos(\psi - \varphi) - \sqrt{3}\sin(\psi - \varphi))}{\sqrt{2}L_F} \right.
$$

$$
\left. \frac{U(\cos(\psi - \varphi) + \sqrt{3}\sin(\psi - \varphi))}{\sqrt{2}L_F}\,000000000 \right).
$$

Taking $\varphi = \psi + \pi/6$ provides

$$B^T = \left(-\sqrt{\frac{3}{2}} \frac{U}{L_F} \sqrt{\frac{3}{2}} \frac{U}{L_F} 0000000000 \right).$$

Solving Equation 2.57 yields

$$Y = e^{(A-\Omega)t} Y_0 + (A-\Omega)^{-1}(e^{(A-\Omega)t} - I)B, \qquad (2.58)$$

where I is the unit matrix; Y_0 is the initial condition vector. From this formula the solution to (2.47) follows at once:

$$X = KY.$$

The steady-state process is obtained from (2.58) as follows:

$$X_{st} = -K(A-\Omega)^{-1}B. \qquad (2.59)$$

The transformation $X = KY$ can also be realized by using matrices:

$$K_S = \sqrt{\frac{2}{3}} \begin{pmatrix} \cos(\omega t) & \sin(\omega t) & \dfrac{1}{\sqrt{2}} \\ \cos\left(\omega t + \dfrac{2\pi}{3}\right) & \sin\left(\omega t + \dfrac{2\pi}{3}\right) & \dfrac{1}{\sqrt{2}} \\ \cos\left(\omega t - \dfrac{2\pi}{3}\right) & \sin\left(\omega t - \dfrac{2\pi}{3}\right) & \dfrac{1}{\sqrt{2}} \end{pmatrix}; \qquad (2.60)$$

$$K_L = \sqrt{\frac{2}{3}} \begin{pmatrix} \cos(\omega_L t) & \sin(\omega_L t) & \dfrac{1}{\sqrt{2}} \\ \cos\left(\omega_L t + \dfrac{2\pi}{3}\right) & \sin\left(\omega_L t + \dfrac{2\pi}{3}\right) & \dfrac{1}{\sqrt{2}} \\ \cos\left(\omega_L t - \dfrac{2\pi}{3}\right) & \sin\left(\omega_L t - \dfrac{2\pi}{3}\right) & \dfrac{1}{\sqrt{2}} \end{pmatrix}.$$

Note that matrices K_S and K_L are inverse with respect to dq-transformation. For this transformation equations (2.56–2.59) are the same, but matrices Ω_S and Ω_L take the forms

$$\Omega_S = \begin{pmatrix} 0 & \omega & 0 \\ -\omega & 0 & 0 \\ 0 & 0 & 0 \end{pmatrix}, \quad \Omega_L = \begin{pmatrix} 0 & \omega_L & 0 \\ -\omega_L & 0 & 0 \\ 0 & 0 & 0 \end{pmatrix}$$

and a part of the matrix A transforms in the following way:

$$
\begin{pmatrix}
\dfrac{a_1}{L_S} & \dfrac{a_2}{L_S} & \dfrac{a_3}{L_S} \\[2ex]
\dfrac{a_3}{L_S} & \dfrac{a_1}{L_S} & \dfrac{a_2}{L_S} \\[2ex]
\dfrac{a_2}{L_S} & \dfrac{a_3}{L_S} & \dfrac{a_1}{L_S}
\end{pmatrix}
\rightarrow
\begin{pmatrix}
\dfrac{q}{L_S} & 0 & 0 \\[2ex]
0 & \dfrac{q}{L_S} & 0 \\[2ex]
0 & 0 & \dfrac{1}{L_S}
\end{pmatrix}
\tag{2.61}
$$

and the vector B^T has the form

$$
B^T = \left(\sqrt{\frac{3}{2}} \frac{U}{L_F} \cos\psi \;-\; \sqrt{\frac{3}{2}} \frac{U}{L_F} \sin\psi \; 0\,0\,0\,0\,0\,0\,0\,0\,0\,0 \right).
$$

When $\psi=0$, the vector B has only the first nonzero component, and the matrix A has more zero components, and then calculations produced by (2.58–2.59) are a little faster.

Let us calculate the transient processes in the matrix-reactance converter. For simplicity we use matrices K_S and K_L in the form (2.60). We start from the beginning. In the first cell we enter the components of the matrix $M(t)$:

$$\omega m = \omega L - \omega;$$

$$daA := Ds*(1+2*qu*Cos[\omega m*t])/3;$$

$$daB := Ds*(1+2*qu*Cos[\omega m*t-2*Pi/3])/3;$$

$$daC := Ds*(1+2*qu*Cos[\omega m*t-4*Pi/3])/3;$$

$$dbA := Ds*(1+2*qu*Cos[\omega m*t-4*Pi/3])/3;$$

$$dbB := Ds*(1+2*qu*Cos[\omega m*t])/3;$$

$$dbC := Ds*(1+2*qu*Cos[\omega m*t-2*Pi/3])/3;$$

$$dcA := Ds*(1+2*qu*Cos[\omega m*t-2*Pi/3])/3;$$

$$dcB := Ds*(1+2*qu*Cos[\omega m*t-4*Pi/3])/3;$$

$$dcC := Ds*(1+2*qu*Cos[\omega m*t])/3;$$

The matrix $M(t)$-denoted **M** is defined in the next cell:

$$
\mathbf{M} =
\begin{pmatrix}
daA & dAB & daC \\
dbA & dbB & dbC \\
dcA & dcB & dcC
\end{pmatrix};
$$

Then we define the matrices:

$$K_S = \sqrt{\frac{2}{3}} * \begin{pmatrix} \text{Cos[} *t] & \text{Sin[} *t] & 1/\sqrt{2} \\ \text{Cos[} *t+2*Pi/3] & \text{Sin[} *t+2*Pi/3] & 1/\sqrt{2} \\ \text{Cos[} *t \ 2*Pi/3] & \text{Sin[} *t \ 2*Pi/3] & 1/\sqrt{2} \end{pmatrix};$$

$$K_L = \sqrt{\frac{2}{3}} * \begin{pmatrix} \text{Cos[} L*t] & \text{Sin[} L*t] & 1/\sqrt{2} \\ \text{Cos[} L*t+2*Pi/3] & \text{Sin[} L*t+2*Pi/3] & 1/\sqrt{2} \\ \text{Cos[} L*t \ 2*Pi/3] & \text{Sin[} L*t \ 2*Pi/3] & 1/\sqrt{2} \end{pmatrix};$$

One can see that, using

MatrixForm[Simplify[KLi.M.KS]]

provides

$$\begin{pmatrix} qu & 0 & 0 \\ 0 & qu & 0 \\ 0 & 0 & 1 \end{pmatrix},$$

which differs from (2.61) only by the factor $1/L_S$. Matrix A denoted by **Aq** is as follows:

Aq =

$$\begin{pmatrix}
\frac{-Rf}{Lf} & 0 & 0 & 0 & 0 & 0 & \frac{-1}{Lf} & 0 & 0 & 0 & 0 & 0 \\
0 & \frac{-Rf}{Lf} & 0 & 0 & 0 & 0 & 0 & \frac{-1}{Lf} & 0 & 0 & 0 & 0 \\
0 & 0 & \frac{-Rf}{Lf} & 0 & 0 & 0 & 0 & 0 & \frac{-1}{Lf} & 0 & 0 & 0 \\
0 & 0 & 0 & \frac{-Rs}{Ls} & 0 & 0 & \frac{q}{Ls} & 0 & 0 & \frac{1-D_S}{Ls} & 0 & 0 \\
0 & 0 & 0 & 0 & \frac{-Rs}{Ls} & 0 & 0 & \frac{q}{Ls} & 0 & 0 & \frac{1-D_S}{Ls} & 0 \\
0 & 0 & 0 & 0 & 0 & \frac{-Rs}{Ls} & 0 & 0 & \frac{1}{Ls} & 0 & 0 & \frac{1-D_S}{Ls} \\
\frac{1}{Cf} & 0 & 0 & \frac{-q}{Cf} & 0 & 0 & 0 & 0 & 0 & 0 & 0 & 0 \\
0 & \frac{1}{Cf} & 0 & 0 & \frac{-q}{Cf} & 0 & 0 & 0 & 0 & 0 & 0 & 0 \\
0 & 0 & \frac{1}{Cf} & 0 & 0 & \frac{-1}{Cf} & 0 & 0 & 0 & 0 & 0 & 0 \\
0 & 0 & 0 & \frac{D_S-1}{Cl} & 0 & 0 & 0 & 0 & 0 & \frac{-1}{RCl} & 0 & 0 \\
0 & 0 & 0 & 0 & \frac{D_S-1}{Cl} & 0 & 0 & 0 & 0 & 0 & \frac{-1}{RCl} & 0 \\
0 & 0 & 0 & 0 & 0 & \frac{D_S-1}{Cl} & 0 & 0 & 0 & 0 & 0 & \frac{-1}{RCl}
\end{pmatrix};$$

and the matrix Ω has the form

$$
= \begin{pmatrix}
0 & 0 & 0 & 0 & 0 & 0 & 0 & 0 & 0 & 0 & 0 \\
- & 0 & 0 & 0 & 0 & 0 & 0 & 0 & 0 & 0 & 0 & 0 \\
0 & 0 & 0 & 0 & 0 & 0 & 0 & 0 & 0 & 0 & 0 & 0 \\
0 & 0 & 0 & 0 & L & 0 & 0 & 0 & 0 & 0 & 0 & 0 \\
0 & 0 & 0 & -L & 0 & 0 & 0 & 0 & 0 & 0 & 0 & 0 \\
0 & 0 & 0 & 0 & 0 & 0 & 0 & 0 & 0 & 0 & 0 & 0 \\
0 & 0 & 0 & 0 & 0 & 0 & 0 & & 0 & 0 & 0 & 0 \\
0 & 0 & 0 & 0 & 0 & 0 & - & 0 & 0 & 0 & 0 & 0 \\
0 & 0 & 0 & 0 & 0 & 0 & 0 & 0 & 0 & 0 & 0 & 0 \\
0 & 0 & 0 & 0 & 0 & 0 & 0 & 0 & 0 & 0 & L & 0 \\
0 & 0 & 0 & 0 & 0 & 0 & 0 & 0 & 0 & -L & 0 & 0 \\
0 & 0 & 0 & 0 & 0 & 0 & 0 & 0 & 0 & 0 & 0 & 0
\end{pmatrix} ;
$$

In the row, one calculates the part of the expression (2.58)

$$A \quad =Aq - \quad ;$$

We define the unit matrix of the twelfth order:

I2=IdentityMatrix[12];

We assume that initial conditions are equal to zero. Then we use the second term in expression (2.58).

Y:=Inverse[A].(MatrixExp[A *t]-I2);

In order to speed up calculations, it is expedient to take into account the structure of the expression

$$K(A-\Omega)^{-1}(e^{(A-\Omega)t}-I)B.$$

The first component of the vector B is not equal to zero. Therefore, we can take only the first column of the matrix

$$(A-\Omega)^{-1}(e^{(A-\Omega)t}-I).$$

Since the matrix K has a diagonal structure, we can multiply only three non-zero elements of K for evaluating one of currents or voltages. In the next cell, functions for currents and voltages are formed:

$$if1:=Part[Y,1,1]*(U/Lf)*Cos[\omega*t]+Part[Y,2,1]*(U/Lf)*Sin[\omega*t]+$$
$$Part[Y,3,1]*(U/Lf)/\sqrt{2};$$

$$is1:=Part[Y,4,1]*(U/Lf)*Cos[\omega L*t]+Part[Y,5,1]*(U/Lf)*Sin[\omega L*t]+$$
$$Part[Y,6,1]*(U/Lf)/\sqrt{2};$$

$$uf1:=Part[Y,7,1]*(U/Lf)*Cos[\omega*t]+Part[Y,8,1]*(U/Lf)*Sin[\omega*t]+$$
$$Part[Y,9,1]*(U/Lf)/\sqrt{2};$$

$$ul1:=Part[Y,10,1]*(U/Lf)*Cos[\omega L*t]+Part[Y,11,1]*(U/Lf)*Sin[\omega L*t]+$$
$$Part[Y,12,1]*(U/Lf)/\sqrt{2};$$

Now we enter the parameters of the circuit elements:

$$Rf=0.01;$$
$$Rs=0.01;$$
$$Lf=0.0005;$$
$$Ls=0.0005;$$
$$Cf=50*10^{\wedge}(-6);$$
$$Cl=50*10^{\wedge}(-6);$$
$$R=10.0;$$
$$U=230.0;$$
$$T=1/50.0;$$
$$\omega=2*Pi/T;$$
$$TL=1/25.0;$$
$$\omega L:=2*Pi/TL;$$
$$qu=0.5;$$
$$Ds=0.7;$$

The graphs of the process (Figures 2.41 and 2.42) are plotted with the help of the **Plot[..]** function

pif1d=Plot[if1,{t,0,2*T},AxesLabel->{"t","I"},DisplayFunction->Identity];

pis1d=Plot[is1,{t,0,2*T},AxesLabel->{"t","I"},DisplayFunction->Identity];

Show[{pis1d,pif1d},DisplayFunction->$DisplayFunction];

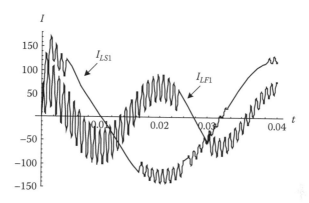

FIGURE 2.41
The transient currents I_{LF1} and I_{LS1} in inductors L_{F1} and L_{S1} (I_{LF1} and I_{LS1} in amperes, time t in seconds).

puf1d=Plot[uf1,{t,0,2*T},AxesLabel->{"t","U"},DisplayFunction ->Identity];

pus1d=Plot[ul1,{t,0,2*T},AxesLabel->{"t","U"},DisplayFunction ->Identity];

Show[{pus1d,puf1d},DisplayFunction->$DisplayFunction];

We could find the steady-state processes by choosing another time interval, for example,

Plot[if1,{t,10*T,12*T},AxesLabel->{"t","I"}];

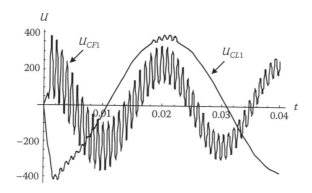

FIGURE 2.42
The transient voltages U_{CF1} and U_{CL1} across capacitors C_{F1} and C_{L1} (U_{CF1} and U_{CL1} in volts, time t in seconds).

Note that the time interval {t,10*T,12*T} for the steady-state process is dependent on a decay rate of the transient process.

We could also verify the obtained results by numerical calculations. In that case we should enter matrix **A(t)** as in (2.54). Then we define the vector of currents and voltages:

$$Xx:=\{i1[t],i2[t],i3[t],i4[t],i5[t],i6[t],u1[t],u2[t],u3[t],u4[t],u5[t],u6[t]\};$$

and form the right part of Equation 2.47:

$$eq:=A11.Xx$$

The components of the vector **Xx** correspond to the components of the vector $X^T = (I_{LF1}\,I_{LF2}\,I_{LF3}\,I_{LS1}\,I_{LS2}\,I_{LS3}\,U_{CF1}\,U_{CF2}\,U_{CF3}\,U_{CL1}\,U_{CL2}\,U_{CL3})$. In the next cell we use the **NDSolve[ ]** function for a numerical solution to Equation 2.47:

sol4Phase = NDSolve{{i1′[t] == eq[[1]] + U * Cos[* t]/Lf, i2′[t] == eq[[2]] +

U * Cos[* t + 2 * Pi/3]/Lf, i3′[t] == eq[[3]] + U * Cos[* t + 4 * Pi/3]/

Lf, i4′[t] == eq[[4]],

i5′[t] == eq[[5]], i6′[t] == eq[[6]], u1′[t] == eq[[7]], u2′[t] == eq[[8]],

u3′[t] == eq[[9]],

u4′[t] == eq[[10]], u5′[t] == eq[[11]], u6′[t] == eq[[12]], i1[0] == 0, i2[0] == 0,

i3[0] == 0,

i4[0] == 0, i5[0] == 0, i6[0] == 0, u1[0] == 0, u2[0] == 0, u3[0] == 0, u4[0] == 0,

u5[0] == 0,

u6[0] == 0}, {i1, i2, i3, i4, i5, i6, u1, u2, u3, u4, u5, u6}, {t, 0, 12 * T},

MaxSteps > 100000];

In order to find a solution in the time interval 0–12T with the required precision, the option **MaxSteps** is set. Using the **Plot[ ]** function

Plot[Part[Evaluate[i1[t]/.sol4Phase],1],{t,0,2*T},AxesLabel->{"t","I"}];

we obtain the same graph for the current I_{LF1}.

3

The Calculation of the Processes and Stability in Closed-Loop Systems

3.1 Calculation of Processes in Closed-Loop Systems with PWM

Electromagnetic processes in converters with a closed-loop feedback are described by nonlinear differential and algebraic equations. For the solution to such equations, numerical and numerical–analytical methods are used. Consider the use of a numerical–analytical method for the calculation of transient and steady-state processes and stability in a Buck-Boost DC voltage converter (Figure 3.1).

Assume that the transistor and diode are described by RS models and in the on state have the same resistances; the inductor and capacitor are linear elements. The control system CS realizes pulse width modulation (PWM) (Korotyeyev and Klytta, 2002). On the inputs of the control system (Figure 3.2), voltages are fed from the load and an independent sawtooth ramp generator. The comparison of the voltages is made on the input of a comparator C. On the output of the comparator, rectangular voltage impulses are formed (Figure 3.3), which open and close the transistor. In the control system, processes are described by the following equation set:

$$u_c = k(u_{ref} - k_r u);$$

$$u_{com} = u_c - u_r; \tag{3.1}$$

$$\gamma = \gamma(u_{com}),$$

where k_r is the output voltage ratio; k is the voltage feedback gain; u_c is the control voltage; u_{ref} is the reference voltage; u_{com} is the voltage on the input of the comparator; u_r is the independent sawtooth ramp voltage; T is the period of the voltage of a generator; τ is the impulse duration on the output of the control system; and $\gamma(t)$ is the switching function (Figure 3.4). The duration

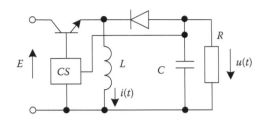

FIGURE 3.1
Topology of a Buck-Boost converter.

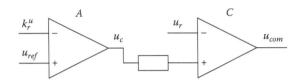

FIGURE 3.2
Control system with PWM-2.

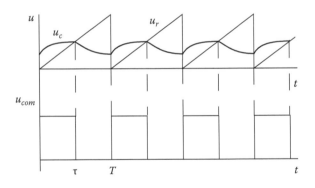

FIGURE 3.3
Time diagrams of the voltages in the control system with PWM.

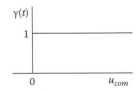

FIGURE 3.4
Switching function.

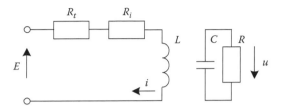

FIGURE 3.5
Equivalent scheme of the converter. The transistor is on, the diode is off.

of the switching function coincides with the duration of the voltage on the output of the comparator.

Let us write differential equations for the intervals of constancy of the converter topology. On the interval $nT \le t \le nT + t_n$ (n is the number of the periods), the transistor is in an on state, and the diode is an off state. The equivalent scheme of the converter is presented in Figure 3.5. The inductor current i and the voltage across the capacitor u are described by the differential equations

$$E = r_1 i + L\frac{di}{dt};$$

$$0 = C\frac{du}{dt} + \frac{u}{R},$$

(3.2)

where $r_1 = r_t + r_i$ is the sum of the resistances of the inductor and the transistor in an on state.

During the interval $nT + t_n \le t \le (n + 1)\, T$, the transistor is off and the diode is on. The equivalent scheme of the converter is presented in Figure 3.6.

The differential equations describing the electromagnetic processes are as follows:

$$0 = r_2 i + L\frac{di}{dt} + u;$$

$$i = C\frac{du}{dt} + \frac{u}{R},$$

(3.3)

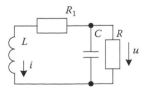

FIGURE 3.6
Equivalent scheme of the converter. The transistor is off, and the diode on.

where r_2 is the sum of resistances of the inductor and the diode in an on state; $r_2 = r_1$.

Let us combine Equations 3.2 and 3.3 using the switching function. We assume that the value of the switching function equal to one corresponds to the on state of the transistor, and the zero value corresponds to the off state of the transistor. Taking this into account, we can write the differential equations for all intervals in the form

$$L\frac{di}{dt} = -r_1 i - (1-\gamma)u + \gamma E; \tag{3.4}$$

$$C\frac{du}{dt} = (1-\gamma)i - \frac{1}{R}u.$$

This equation set we represent as follows:

$$\frac{dX(t)}{dt} = A(\gamma)X(t) + B(\gamma), \tag{3.5}$$

where

$$X(t) = \begin{vmatrix} i \\ u \end{vmatrix}$$

is the vector of the state variables;

$$A(\gamma) = \begin{vmatrix} -\dfrac{r_1}{L} & -\dfrac{1-\gamma}{L} \\ \dfrac{1-\gamma}{C} & -\dfrac{1}{RC} \end{vmatrix}; \quad B(\gamma) = \begin{vmatrix} \dfrac{\gamma E}{L} \\ 0 \end{vmatrix}.$$

The set of Equations 3.1–3.5 describes processes in the closed-loop system of the converter with PWM.

Let us consider the use of Mathematica for discovering transient behaviors. In the next cell, the variables are defined and values assigned to them

R1=0.05;

L1=40*10^(-6);

C1=2.0*10^(-6);

Rn=10.0;

E1=12;

T=10.0*10^(-6);

$$Kd=0.01;$$

$$Ky=1.6;$$

$$A1 = \begin{pmatrix} -R1/L1 & 0 \\ 0 & -1/(Rn * C1) \end{pmatrix};$$

$$A2 = \begin{pmatrix} -R1/L1 & -1/L1 \\ 1/C1 & -1/(Rn * C1) \end{pmatrix};$$

$$B1=\{E1/L1, 0\};$$

$$Ii=\{0, 1\};$$

$$I2=IdentityMatrix[2];$$

$$Ug=5.0;$$

$$Uref=1.5;.$$

$$Kg=Ug/T;$$

In this cell, **Rn** is the load resistance; **Ky** is the voltage feedback gain k; **A1, A2** are the matrixes coinciding with the matrix $A(\gamma)$ for $\gamma = 1$ and $\gamma = 0$, respectively; **B1** is the vector B for $\gamma = 1$; **Ii** is the vector that extracts the second component (it is necessary for extracting the voltage from the vector $X(t)$); **Uref** is the reference voltage; and **Ug** is the voltage amplitude of the independent generator.

In the next cell, solutions for all intervals of constancy of converter topologies and the expression of transitional process are presented, and a solution for a nonlinear algebraic equation is executed. For the interval $nT \le t \le nT + t_n$, when the transistor is on the processes are described by the expression

$$X(t) = e^{A1(t-nT)} X(nT) + A_1^{-1} (e^{A1(t-nT)} - I)B. \tag{3.6}$$

For the interval when the transistor is off, the processes are described as

$$X(t) = e^{A2(t-nT-t_n)} X(nT + t_n). \tag{3.7}$$

Substituting $t = nT + t_n$ in (3.6) and $t = (n + 1)T$ in (3.7), and taking into account the periodicity condition $X(nT) = X((n + 1)T)$, we find the steady-state process:

$$X(nT) = (I - e^{A2(T-t_n)} e^{A1t_n})^{-1} \left[e^{A2(T-t_n)} A_1^{-1} (e^{A1t_n} - I) \right]B. \tag{3.8}$$

In this expression $t_n = const.$

In the cell

Clear[tn];

A1inv=Inverse[A1];

A2inv=Inverse[A2];

An1:=MatrixExp[A1*tn];

An2:=MatrixExp[A2*(T-tn)];

ATn:=An2.An1;

ATninv:=Inverse[I2-ATn];

XTn:=ATninv.(An2.A1inv.(An1-I2)).B1;

Xn1:=An1.XTn+A1inv.(An1-I2).B1;

F:=Kg*tn-Ky*(Uref-Chop[Kd*(Ii.Xn1)])   (3.9)

Ftn=FindRoot[F==0,{tn,T/2,0,T}];

tu=Chop[tn/.Ftn[[1]]]

tn=tu;

XTn corresponds to the vector $X(nT)$, **Xn1** corresponds to the vector (3.6) for $t = nT + t_n$. The function **F** is defined by the second equation of the set (3.1), in which the first equation and value of the voltage from (3.6) are substituted:

$$F = -u_r + k(u_{ref} - k_r u).\tag{3.10}$$

The variable **Ftn** denotes the result of the solution to the nonlinear algebraic equation. The solution is found with the help of the **FindRoot[ ]** function. In the list **{tn,T/2,0,T}**, the variable with respect to which a solution is found, the initial point, and the interval for which it is necessary to find the solution are defined. As a result, we determine the value of time t_n:

4.44179×10^{-6}

Let us plot the time diagrams of the steady-state processes. In the cell

XT=XTn;

Xt1=Xn1;

Y1[t_]:=If[t<t1,MatrixExp[A1*t].XT+A1inv.(MatrixExp[A1*t]-I2).B1,

MatrixExp[A2*(t-t1)].Xt1];

Plot[Part[Y1[u],1],{u,0,T},AxesLabel->{"t","i"}];

Plot[Part[Y1[u],2],{u,0,T},AxesLabel->{"t","u"}];

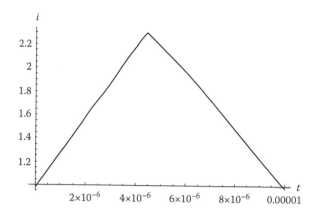

FIGURE 3.7
Steady-state inductor current (i in amperes, time t in seconds).

with the help of the **If[ ]** function for two intervals $0 \le t < t_1$ and $t_1 \le t \le T$, two solutions with initial conditions **XT=XTn** and **Xt1=Xn1** are combined. This solution corresponds to a steady-state process. The graphs of the process (Figures 3.7 and 3.8) are plotted with the help of the **Plot[ ]** function.

Let us calculate the transient process in the circuit of the converter. Assume that the inductor current is positive. In that case, it is not necessary to carry out a corresponding check or find the time at which the current equals zero. Solving this is based on the solution to Equation 3.10 during an on state of the transistor and on recurrent use of the relations (3.6) and (3.7). The initial value of the vector is set equal to zero.

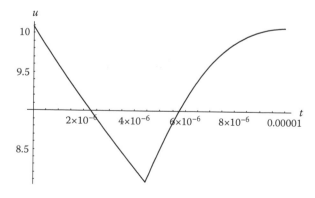

FIGURE 3.8
Steady-state voltage across the capacitor (u in volts, time t in seconds).

In the cell

Clear[tn];

Kp=20;

X0={0,0};

Tg=0.85*T;

Xn1:=An1.X0+A1inv.(An1-I2).B1;

X1s[1]=X0;

For[n=1,n≤Kp,n++,Ftn=FindRoot[F==0,{tn,T/2,0,T}];

t1=Re[tn/.Ftn[[1]]];If[t1>Tg,tn=Tg,tn=t1];Xt[n]=tn;

X2s[n]=Xn1;X0=An2.Xn1;X1s[n+1]=X0]

the **Kp** variable defines the number of periods for which the transient process is calculated. The vector **X0 = {0,0}** defines the initial value of the vector of space variables. The variable **Tg=0.85*T** defines the maximum value of on-state time for the transistor. The introduction of this variable is governed by a static characteristic of the converter with a closed-loop control system. The static characteristic is defined by the expression

$$\frac{u}{E} = \frac{\gamma(1-\gamma)}{[r_1/R-(1-\gamma)^2]}$$

and has the part with a negative derivative (Figure 3.9). This characteristic is obtained by setting the right part of Equation 3.5 equal to zero.

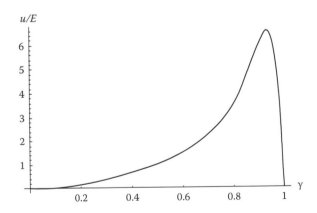

FIGURE 3.9
The static characteristic of the converter.

With the help of the **For[n=1,n≤Kp,n++...]** function, calculations are made for the moments of crossing of the control u_c and the generator u_r voltages, and for the values of the vectors $X(nT)$ and $X(nT + t_n)$. These values are set in the arrays **Xt[n]**, **X1s[n]**, and **X2s[n]**, respectively.

To plot the time diagram, the **If[ ]** function is used to combine the two interval solutions obtained earlier:

Y2[t_]:=If[(t>=(Floor[t/T])*T)&&(t<(Floor[t/T])*T+Xt[Floor[t/T]+1]),

MatrixExp[A1*(t-(Floor[t/T])*T)].X1s[(Floor[t/T]+1)]+

A1inv.(MatrixExp[A1*(t-(Floor[t/T])*T)]-I2).B1,

MatrixExp[A2*(t-(Floor[t/T])*T-Xt[Floor[t/T]+1])].X2s[Floor[t/T]+1]];

Plot[Part[Y2[t],1],{t,0,20*T},AxesLabel->{"t","i"}];

Plot[Part[Y2[t],2],{t,0,20*T},AxesLabel->{"t","u"}];

The **Floor[t/T]** function is used to determine the number of the period. The expression

(t>=(Floor[t/T])*T)&&(t<(Floor[t/T])*T+Xt[Floor[t/T]+1])

corresponds to the interval $nT \le t < nT + t_n$. Time diagrams of the transient process for the current and voltage are presented in Figures 3.10 and 3.11, respectively.

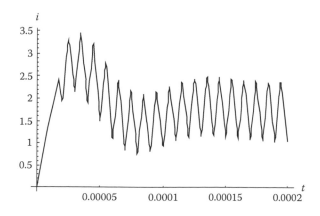

FIGURE 3.10

Time diagram of the transient inductor current (i in amperes, time t in seconds).

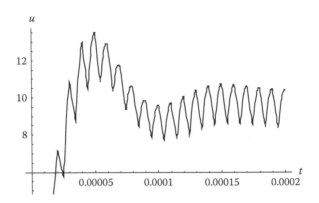

FIGURE 3.11
Time diagram of the transient voltage across the capacitor (*u* in volts, time *t* in seconds).

It can be seen from Figure 3.11 that Mathematica® plots only part of the time diagram. In order for the whole time diagram to be plotted, we use the **PlotRange->{0,16}** option, that is,

Plot[Part[Y2[t],2],{t,0,20*T},AxesLabel->{"t","u"},PlotRange->{0,16}];

The whole time diagram of the transient voltage across the capacitor is shown in Figure 3.12. From the diagrams one sees overshot and large ripples in the voltage.

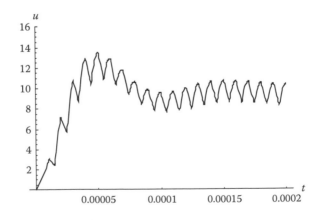

FIGURE 3.12
Time diagram of the transient voltage across the capacitor (*u* in volts, time *t* in seconds). The diagram is plotted with the use the **PlotRange->{0,16}** option.

3.2 Stability Analysis in Closed-Loop Systems with PWM

The equation set (3.1) and (3.5) is nonlinear. A stability analysis will be based upon the first Lyapunov method. For the stability analysis we use techniques (Tsypkin, 1974; Rozenwasser and Yusupov, 1981) that permit the execution of linearization of a differential and algebraic equation set described through intervals of constancy topology. In contrast to this method, the method presented in Bromberg (1967) requires prior solution to a nonlinear differential equation set.

Let us linearize Equations 3.1 and 3.2 around a steady-state process (Zhuykov et al., 1989). We vary the state space variable corresponding to initial conditions on the infinitesimal value $X_\xi(mT)$. We find equations describing changes of the state space variables for time moments $t > mT$. Since solutions depend on initial values and time, that is, $X(t) = f(X(mT), t)$, then the increment of state space variables is determined by finding the variation

$$X_\xi = \frac{\partial f}{\partial X(mT)} X_\xi(mT). \tag{3.11}$$

Applying (3.11) to set (3.1) and (3.5) yields

$$\frac{dX_\xi}{dt} = A_\xi(\gamma)X + A(\gamma)X_\xi + B_\xi(\gamma); \tag{3.12}$$

$$u_{\xi c} = -kk_r u_\xi;$$

$$u_{\xi com} = u_{\xi c}.$$

Substituting for $u_{\xi com}$, we obtain

$$u_{\xi com} = -kk_r u_\xi.$$

Let us determine $A_\xi(\gamma)$ and $B_\gamma(\gamma)$. Since the matrix $A(\gamma)$ depends on the voltage u, $A_\xi(\gamma) = \frac{\partial A(\gamma)}{\partial \gamma} \frac{\partial \gamma}{\partial u_{com}} \frac{\partial u_{com}}{\partial u} u_\xi$. In this expression, $\frac{\partial u_{com}}{\partial u} = -kk_r$.

Let us determine the derivative $\frac{\partial \gamma}{\partial u_{com}}$. Since the switching function γ is the step function, the derivative $\frac{\partial \gamma}{\partial u_{com}} = \delta(u_{com})$ (δ being the Dirac delta function). The switching function depends also on time $\gamma = \gamma(u_{com}(t))$ (Figure 3.13). The derivative with respect to time is

$$\frac{\partial \gamma}{\partial t} = \sum_{\mu=0}^{\infty} (-1)^\mu \delta(t - t_\mu), \tag{3.13}$$

where t_μ are time moments in which control impulse durations are changed with the changing of the initial values.

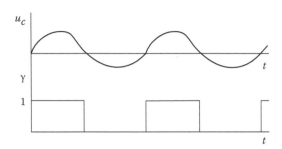

FIGURE 3.13
Relation between the control voltage and moments of change of the switching function.

Calculating the derivative of the switching function as a composite function, we obtain

$$\frac{\partial \gamma(u_{com}(t))}{\partial t} = \frac{\partial \gamma}{\partial u_{com}} \frac{du_{com}}{dt}, \tag{3.14}$$

at which the derivative is equal to $\frac{\partial \gamma}{\partial u_{com}} = \delta(u_{com})$.

We denote the derivative as $\frac{du_{com}}{dt} = u_{tcom}$. Using expressions (3.13) and (3.14), we obtain

$$\delta(u_{com})u_{tcom} = \sum_{\mu=0}^{\infty} (-1)^{\mu} \delta(t - t_{\mu}).$$

Then,

$$\delta(u_{com}) = \frac{\displaystyle\sum_{\mu=0}^{\infty} (-1)^{\mu} \delta(t - t_{\mu})}{u_{tcom}}. \tag{3.15}$$

The derivative u_{tcom} is a positive in the case $(-1)^{\mu}(t-t_{\mu}) > 0$; therefore,

$$\delta(u_{com}) = \frac{\displaystyle\sum_{\mu=0}^{\infty} \delta(t - t_{\mu})}{|u_{tcom}|}.$$

Taking into account that, for any continuous function $y(t)$, the product $y(t)\delta(t-t_{\mu}) = y(t_{\mu})\delta(t-t_{\mu})$, the expression (3.15) becomes

$$\delta(u_{com}) = \frac{\displaystyle\sum_{\mu=0}^{\infty} \delta(t - t_{\mu})}{|u_{tcom}(t_{\mu})|},$$

where $u_{tcom}(t_{\mu}) = \lim\limits_{t \to t_{\mu}-0} \frac{du_{com}(t)}{dt}$ is the derivative on the left of the function $u_{com}(t)$ at the point t_{μ}. This implies that a control system at first states the value of the control signal, and then the switching is made.

Using the obtained expressions, we find that

$$A_\xi(\gamma) = -kk_r A_\gamma \sum_{\mu=0}^{\infty} \frac{\delta(t-t_\mu)}{|u_{tcom}(t_\mu)|} u_\xi(t_\mu),$$

where $A_\gamma = \frac{\partial A}{\partial \gamma}$. Similarly,

$$B_\xi(\gamma) = -kk_r B_\gamma \sum_{\mu=0}^{\infty} \frac{\delta(t-t_\mu)}{|u_{tcom}(t_\mu)|} u_\xi(t_\mu).$$

Substituting the obtained expressions in (3.12), we have

$$\frac{dX_\xi}{dt} = A(\gamma)X_\xi - kk_r \sum_{\mu=0}^{\infty} \frac{B_\gamma(\gamma) + A_\gamma(\gamma)X(t_\mu)}{|u_{tcom}(t_\mu)|} u_\xi(t_\mu)\delta(t-t_\mu). \qquad (3.16)$$

Equation 3.16 is a linear nonstationary differential equation. In order to determine stability conditions, it is necessary to find the solution to this equation for the period T. Denoting

$$D_\mu = -kk_r \frac{B_\gamma(\gamma) + A_\gamma(\gamma)X(t_\mu)}{|u_{tcom}(t_\mu)|},$$

we write Equation 3.16 in the form

$$\frac{dX_\xi}{dt} = A(\gamma)X_\xi + \sum_{\mu=0}^{\infty} D_\mu u_\xi(t_\mu)\delta(t-t_\mu), \qquad (3.17)$$

where

$$A_\gamma(\gamma) = \begin{vmatrix} 0 & \dfrac{1}{L} \\ -\dfrac{1}{C} & 0 \end{vmatrix}; \quad B_\gamma(\gamma) = \begin{vmatrix} \dfrac{E}{L} \\ 0 \end{vmatrix}.$$

In order to calculate the derivative $u_{tcom}(t_\mu)$ at the point t_μ, it is necessary to determine a steady-state process in the closed-loop system.

Let us find stability conditions for the steady-state process. For this we will find the solution to Equation 3.17 for the part equal to the period of a generator voltage. For the first interval $mT \le t \le mT + \tau$, the differential equation (3.16) has the form

$$\frac{dX_\xi}{dt} = A_1 X_\xi + D_1 u_\xi(mT)\delta(t-mT), \qquad (3.18)$$

where

$$A_1 = A(\gamma)|_{\gamma=1}; \quad A_1 = \begin{vmatrix} -\dfrac{r}{L} & 0 \\ 0 & -\dfrac{1}{RC} \end{vmatrix}; \quad D_1 = -\begin{vmatrix} \dfrac{kk_r}{k_1} & \dfrac{u(mT)+E}{L} \\ & \dfrac{i(mT)}{C} \end{vmatrix}; \quad k_1 = |u_{tcom}(mT)|.$$

The value $X(mT)$ is not dependent on the number of the period (a steady-state process is considered). Therefore, the vector D_1 is constant.

Applying the Laplace transform, we will find the solution to Equation 3.18. The transformation of Equation 3.18, taking into account the initial condition $X_\xi(mT)$, has the form

$$pX_\xi(p) = A_1 X_\xi(p) + [D_1 u_\xi(mT) + X_\xi(mT)]e^{-pmT}, \tag{3.19}$$

where $X_\xi(p)$ is the transformation of the vector X_ξ.

Transforming the expression in square brackets in Equation 3.19, we obtain

$$pX_\xi(p) = A_1 X_\xi(p) + N_1 X_\xi(mT)e^{-pmT},$$

where $D_1 u_\xi(mT) + X_\xi(mT) = N_1 X_\xi(mT)$;

$$D_1 = \begin{vmatrix} d_1^1 \\ d_1^2 \end{vmatrix}; \quad N_1 = \begin{vmatrix} 1 & d_1^1 \\ 0 & 1+d_1^2 \end{vmatrix}; \quad d_1^1 = -\frac{kk_r[u(mT)+E]}{k_1 L}; \quad d_1^2 = -\frac{kk_r i(mT)}{k_1 C}.$$

Solving this equation yields

$$X_\xi(p) = (pI - A_1)^{-1} N_1 X_\xi(mT)e^{-pmT}.$$

Taking into account that the original of the transformation $(pI - A_1)^{-1}e^{-pmT}$ is the matrix exponential $e^{A_1(t-mT)}$, we find the original for the transformation $X_\xi(p)$. Then,

$$X_\xi(t) = e^{A_1(t-mT)} N_1 X_\xi(mT).$$

Substituting in this expression the value of time $t = mT + \tau$ equal to the end of the interval yields

$$X_\xi(mT+\tau) = e^{A_1\tau} N_1 X_\xi(mT). \tag{3.20}$$

For the considered generator voltage (see Figure 3.3), the descending part is vertical. Therefore, $k_1 \to \infty$, and so $N_1 = I$.

For the second part of the interval of constancy topology of converter $mT + \tau \le t \le (m+1)T$ $\gamma = 0$, Equation 3.17 is

$$\frac{dX_\xi}{dt} = A_2 X_\xi + D_2 u_\xi (mT + \tau), \tag{3.21}$$

where

$$D_2 = -kk_r \frac{B_\gamma(\gamma) + A_\gamma(\gamma) X(mT + \tau)}{|u_{tcom}(mT + \tau)|} = -\frac{kk_r}{k_2} \begin{vmatrix} \dfrac{u(mT + \tau)}{L} \\ -\dfrac{i(mT + \tau)}{C} \end{vmatrix}; \quad k_2 = |u_{tcom}(mT + \tau)|.$$

The solution to the differential equation (3.21) is determined similarly to the solution to Equation 3.18. Applying the Laplace transform, we express the solution to Equation 3.21 in the form

$$X_\xi(p) = (pI - A_2)^{-1} [D_2 u_\xi(mT + \tau) + X_\xi(mT + \tau)] e^{-(pmT+\tau)}, \tag{3.22}$$

where $X_\xi(mT + \tau)$ is the initial value of the vector X_ξ for the moment $t = mT + \tau$.

We define a matrix N_2 as follows

$$D_2 u_\xi(mT + \tau) + X_\xi(mT + \tau) = N_2 X_\xi(mT + \tau),$$

where

$$D_2 = \begin{vmatrix} d_2^1 \\ d_2^2 \end{vmatrix}; \quad N_2 = \begin{vmatrix} 1 & d_2^1 \\ 0 & 1 + d_2^2 \end{vmatrix}; \quad d_2^1 = -\frac{kk_r u(mT + \tau)}{k_2 L}; \quad d_2^2 = -\frac{kk_r i(mT + \tau)}{k_2 C}.$$

Using the matrix N_2, we write (3.22) in the form

$$X_\xi(p) = (pI - A_2)^{-1} N_2 X_\xi(mT + \tau) e^{-(pmT+\tau)}.$$

Taking into account that the original of the transformation $(pI - A_2)^{-1} e^{-(pmT+\tau)}$ is the matrix exponential, the solution takes the form

$$X_\xi(t) = e^{A_2(t-mT-\tau)} N_2 X_\xi(mT + \tau),$$

where $e^{A_2(t-mT-\tau)}$ is the matrix exponential.

Substituting in this expression the value of time corresponding to the end of the period, that is, $t = (m + 1)T$, yields

$$X_\xi((m+1)T) = e^{A_2(T-\tau)}N_2 X_\xi(mT+\tau). \tag{3.23}$$

Next, substituting in (3.23) the value $X_\xi(mT + \tau)$ from (3.20), we obtain a difference equation

$$X_\xi((m+1)T) = e^{A_2(T-\tau)}N_2 e^{A_1\tau}N_1 X_\xi(mT). \tag{3.24}$$

In line with Lyapunov's first method, if the solution to Equation 3.24 is stable, the initial nonlinear system is stable (stability in small).

The stability of the linearized system is determined by the eigenvalues of the matrix:

$$H = e^{A_2(T-\tau)}N_2 e^{A_1\tau}N_1 \tag{3.25}$$

The system will be stable if all absolute values of the eigenvalues of the matrix H will be less then unity.

According to the form of generator voltage, the matrix H takes a different form. The expression (3.25) corresponds to the case when both front edges of the generator voltage have finite slopes. For the case under consideration (see Figure 3.3), the expression takes the form

$$H = e^{A_2(T-\tau)}N_2 e^{A_1\tau}. \tag{3.26}$$

For the computation of the eigenvalues it is necessary to find elements of matrixes N_1 and N_2. At first the values of the vectors of the state variables $X(mT)$, $X(mT + \tau)$ for the steady-state process should be determined. Next, the values of coefficients k_1 and k_2 should be computed. For the calculation of the derivative, we use the initial differential equation set

$$\frac{dX}{dt} = A_1 X + B_1, \quad mT \le t \le mT + \tau;$$

$$\frac{dX}{dt} = A_2 X + B_2, \quad mT + \tau \le t \le (m+1)T.$$

The left-hand-side derivatives of the vector X at the moments of structure changing are defined by the expressions

$$\left.\frac{dX}{dt}\right|_{t=mT+\tau-0} = A_1 X(mT+\tau) + B_1; \tag{3.27}$$

$$\left.\frac{dX}{dt}\right|_{t=mT-0} = A_2 X(mT) + B_2.$$

The elements of (3.27) are the derivatives

$$u_t(mT+\tau) = \frac{du}{dt}\bigg|_{t-mT+\tau-0} \quad , \quad u_t(mT) = \frac{du}{dt}\bigg|_{t=mT-0} \quad ,$$

which are used for computation of the values $u_{tcom}(mT+\tau) = -kk_r u_t(mT+\tau)$ and $u_{tcom}(mT) = -kk_r u_t(mT)$, respectively.

Let us consider the use of Mathematica for stability analysis. We continue with the analysis of the Buck-Boost converter. In the cell, the eigenvalues of the matrix H are computed:

Udt2=Ii.(A1.Xt1+B1);

D2=-((A1-A2).Xt1+B1)*Kd*Ky/Abs[-Kg-Ky*Kd*Udt2];

$$N2 = \begin{pmatrix} 1.0 & D2[[1]] \\ 0 & 1+D2[[2]] \end{pmatrix};$$

At1=MatrixExp[A1*tu];

At2=MatrixExp[A2*(T-tu)];

Hs=At2.N2.At1;

Sei=Eigenvalues[Hs]

Abs[Sei]

The variable **Udt2** corresponds to the derivative $u_t(mT+\tau)$, **Hs** denotes the matrix H, **tu** denotes τ, and **Sei** denotes the eigenvalues of the matrix H. As a result of computations, we determine the matrix H eigenvalues

$$\{0.644784 + 0.454581i, \quad 0.644784 - 0.454581i\}$$

and the absolute value of the eigenvalues $\{0.788283, 0.788283\}$. Since the absolute value is less then unity, the system is stable.

The calculations show that the increase in the voltage feedback gain k leads to increase in the absolute values of the eigenvalues. For $k = 3.45$, the absolute values are $(0.997657, 0.997657\}$, and for $k = 3.46$, the absolute values are $\{1.0003, 1.0003\}$. Therefore, for $k = 3.46$, the system becomes unstable.

It should be noted that, for $k = 3.46$, the accuracy of calculation of the switching moment becomes unsatisfactory. The error of the calculation of the voltage defined by the expression (3.9) is equal to -0.27839. At that the duty factor equals 0.774163. To improve accuracy, it is necessary to change the method of calculation of pulse duration. Replace expression

Ftn=FindRoot[F==0,{tn,T/2,0,T}];

with the sequence

$$delta=0.0001;$$

$$ta=0.01*T;$$

$$tb=0.9*T;$$

$$While[(tb-ta)/T>delta,\{tn=(ta+tb)/2;If[F>0,tb=tn,ta=tn]\}];$$

This determines the dichotomy method. The **delta** variable defines the accuracy of computation of impulse durations, and **ta** and **tb** are the beginning and end of the time interval on the period **T**. In that case, F = 0.000120373, and the duty factor equals 0.801081. Taking into account this value, we determine the absolute values of eigenvalues {1.05969, 1.05969}.

Changing the form of generator voltage leads to changes in the stability of the system. For the declining form of generator voltage (Figure 3.14), the matrix H is defined by the expression

$$H = e^{A_2(T-\tau)}e^{A_1\tau}N_1. \tag{3.27b}$$

By changing the equation (3.9) for

$$F:=Kg*tn-Ky*(Uref-Chop[Kd*(Ii.XTn)]);$$

and the expressions of the derivative $u_t(mT)$, vector D_1, matrix N_1 for

$$Udt1=Ii.(A2.XT);$$

$$D1=-((A1-A2).XT+B1)*Kd*Ky/Abs[-Kg-Ky*Kd*Udt1];$$

$$N1 = \begin{pmatrix} 1.0 & D1[[1]] \\ 0 & 1+D1[[2]] \end{pmatrix};$$

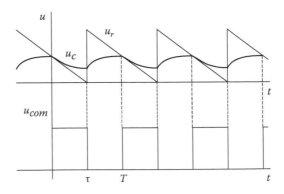

FIGURE 3.14
Time diagrams of the signals in the control system with PWM.

$$At1 = MatrixExp[A1*tu];$$

$$At2 = MatrixExp[A2*(T-tu)];$$

$$Hs = At2.At1.N1;$$

$$Sei = Eigenvalues[Hs]$$

$$Abs[Sei]$$

we determine the matrix H eigenvalues

$$\{0.863092 + 0.486529i, \quad 0.863092 - 0.486529i\}$$

and the absolute values of eigenvalues $\{0.990777, 0.990777\}$ for $k = 3.46$ (the switching moment is computed by the dichotomy method).

3.3 Stability Analysis in Closed-Loop Systems with PWM Using the State Space Averaging Method

We consider the use of the state space averaging method (Middlebrook and Ćuk, 1976) for stability analysis in the Boost converter (Figure 2.8) with a closed-loop system. Processes in such a converter are described by the differential equation

$$\frac{dX(t)}{dt} = A(\gamma)X(t) + B(\gamma), \tag{3.28}$$

where

$$X(t) = \begin{vmatrix} i \\ u \end{vmatrix}$$

is the vector of the state variables;

$$A(\gamma) = \begin{vmatrix} -\dfrac{r_1}{L} & -\dfrac{1-\gamma}{L} \\ \dfrac{1-\gamma}{C} & -\dfrac{1}{RC} \end{vmatrix} ; \quad B(\gamma) = \begin{vmatrix} \dfrac{E}{L} \\ 0 \end{vmatrix} ;$$

γ is the switching function (Figure 3.4). The control system processes are described by the following equation set:

$$u_c = k(u_{ref} - k_r u);$$

$$u_{com} = u_c - u_r ; \tag{3.29}$$

$$\gamma = \chi(u_{com}),$$

where k_r is the output voltage ratio; k is the voltage feedback gain; u_c is the control voltage; u_{ref} is the reference voltage; u_{com} is the voltage on the input of the comparator; and u_r is the independent sawtooth ramp voltage.

Solution to Equation 3.28 for $\gamma = 1$ is

$$X(t) = e^{A_1(t-mT)}X(mT) + (A_1)^{-1}(e^{A_1(t-mT)} - I)B_1 \tag{3.30}$$

and for $\gamma = 0$ is

$$X(t) = e^{A_2(t-mT-t_m)}X(mT+t_m) + (A_2)^{-1}(e^{A_2(t-mT-t_m)} - I)B_2 \tag{3.31}$$

where t_m is the value of the time when the topology of the converter is changed; and T is the period of an independent generator.

Substituting in (3.30) the value $t = mT + t_m$, and in (3.31) the value $t = (m + 1)T$ and eliminating $X = (mT + t_m)$, we get

$$X((m+1)T) = e^{A_2(T-t_m)}e^{A_1t_m}X(mT) + Q, \tag{3.32}$$

where $Q = e^{A_2(T-t_m)}((A_1)^{-1}(e^{A_1t_m} - I)B_1) + (A_2)^{-1}(e^{A_2(T-t_m)} - I)B_2$. Using the periodicity condition $X((m+1)T) = X(mT)$, one obtains from (3.32) the initial value for a steady-state process:

$$\hat{X}(0) = (I - e^{A_2(T-\tau)}e^{A_1\tau})^{-1}Q \tag{3.33}$$

where τ denotes the value of t_m for the steady-state process. We determine the value of τ as a result of solving the nonlinear algebraic equation

$$\frac{U_g}{T}\tau = k[u_{ref} - k_r\hat{u}(\tau)], \tag{3.34}$$

where $\hat{u}(\tau)$ is the voltage for the steady-state process. This voltage is obtained from (3.30) by substituting $\hat{X}(0)$ in this equation.

Using the state space averaging method, we transform Equations 3.28 and 3.29 into

$$\frac{d\bar{X}}{dt} = A(d)\bar{X} + B(d);$$
$$U_gd = k(u_{ref} - k_r\bar{u}(d)), \tag{3.35}$$

where $\bar{X} = \begin{vmatrix} \bar{i} \\ \bar{u} \end{vmatrix}$ is the vector of averaged state variables; $A(d) = A_1d + A_2(1-d)$;

$B(d) = B_1d + B_2(1-d)$; d is the averaged value of the switching function γ on the period T; and

$$A_1 = \begin{pmatrix} -\frac{r}{L} & 0 \\ 0 & -\frac{1}{RC} \end{pmatrix}; \quad A_2 = \begin{pmatrix} -\frac{r}{L} & -\frac{1}{L} \\ \frac{1}{C} & -\frac{1}{RC} \end{pmatrix}; \quad B_1 = B_2 = \begin{pmatrix} \frac{E}{L} \\ 0 \end{pmatrix}.$$

In order to find the steady-state value of the vector $\hat{\bar{X}}$, we equate to zero the right part of the equation (3.35)

$$\hat{\bar{X}} = -A(D)^{-1}B,$$

(3.36)

or

$$\begin{pmatrix} \hat{\bar{i}} \\ \hat{\bar{u}} \end{pmatrix} = \begin{pmatrix} \dfrac{E}{r+(1-D)^2 R} \\[2mm] \dfrac{(1-D)RE}{r+(1-D)^2 R} \end{pmatrix},$$

where D is the value of d for the steady-state process; and $\hat{\bar{X}}$ is the steady-state value of $\bar{X}$. The second equation (3.35) in the steady state has the form

$$U_g D = k(u_{ref} - k_r \hat{\bar{u}}(D))$$

(3.37)

We can find the steady-state value of the vector $\hat{\bar{X}}$ and D by simultaneous calculation equations (3.36) and (3.37):

$$(U_g D - k u_{ref})(r+(1-D)^2 R) + kk_r(1-D)RE = 0.$$

Let us linearize the equation set (3.35) around the steady-state process $\hat{\bar{X}}$. Then,

$$\frac{d\bar{X}_\xi}{dt} = A(D)\bar{X}_\xi + [A_d(D)\hat{\bar{X}} + B_d(D))]d_\xi;$$

(3.38)

$$d_\xi = -k_e \bar{u}_\xi,$$

where $\bar{X}_\xi$ is the increment of the averaged state vector $\bar{X}$; $k_e = \frac{kk_r}{U_g}$; $A_d(D) = \frac{\partial A(d)}{\partial d}\Big|_{d=D}$; and $B_d(D) = \frac{\partial B(d)}{\partial d}\Big|_{d=D}$.

Substituting the second expression of (3.38) in the first yields

$$\frac{d\bar{X}_\xi}{dt} = FX_\xi,$$

(3.39)

$$F = A(D) - k_e[A_d(D)\hat{\bar{X}} + B_d(D)]G.$$

(3.40)

where $G_{\wedge} = |01|$ is the vector with two elements. Since the expression $k_e[A_d(D)\hat{\bar{X}} + B_d(D)]$ has the structure of a vector column, and G is the vector row, their multiplication will be a matrix with the first column equal to zero.

The system (3.39) is stable when all real parts of the eigenvalues of the matrix F are negative.

Let us compare stability conditions obtained by the averaged and exact methods. In the cell we input parameter values

$$r=0.005;$$

$$L=40.0*10^{\wedge}(-6);$$

$$C1=1.0*10^{\wedge}(-6);$$

$$Rn=20.0;$$

$$E1=4;$$

$$T=25.0*10^{\wedge}(-6);$$

$$t2:=T-t1;$$

$$Kd=0.01;$$

$$Ky=1.49;$$

$$A_1 = \begin{pmatrix} -\dfrac{r}{L} & 0 \\ 0 & -\dfrac{1}{RC} \end{pmatrix}$$

$$A_2 = \begin{pmatrix} \dfrac{-r}{L} & -\dfrac{1}{L} \\ \dfrac{1}{C} & -\dfrac{1}{RC} \end{pmatrix}$$

$$Ev=\{E1/L,0\};$$

$$I2=IdentityMatrix[2];$$

$$Ii=\{0,1\};$$

$$Ug=1.0;$$

$$Uref=0.6;$$

$$Kg=Ug/T;$$

$$Dd:=t1/T;$$

In this cell, **Rn** denotes the resistance R; **Ky** denotes the voltage feedback gain k; **B1** denotes the vector B; **Kd** denotes the output voltage ratio k_r; and **Ii** defines the vector that extracts the second component.

To begin, we find the steady-state solution for the nonlinear set of equations (3.28) and (3.29). For this we solve Equation 3.34 together with Equation 3.33

A1inv=Inverse[A1];

A2inv=Inverse[A2];

An1:=MatrixExp[A1*tn];

An2:=MatrixExp[A2*(T-tn)];

ATn:=An2.An1;

ATninv:=Inverse[I2-ATn];

XTn:=ATninv.(An2.A1inv.(An1-I2)+A2inv.(An2-I2)).Ev;

Xn1:=An1.XTn+A1inv.(An1-I2).Ev;

F:=Kg*tn-Ky*(Uref-Kd*(Ii.Xn1));

Ftn=FindRoot[F==0,{tn,T/2,0,T}]

t1=Re[tn/.Ftn[[1]]];

In this cell, **XTn** corresponds to $X(mT) = X(0)$, and **Xn1** corresponds to the vector (3.6) for $t = nT + t_m$. The function **F** is defined by the second equation of the set (3.34). As a result, we determine the value of time t_n:

$$\{t_n - > 0.0000198723 + 5.07691 \times 10^{-25} i\}$$

Now we form the matrix H (3.26) and calculate its eigenvalues:

Udt2=Ii.(A1.Xt1);

D2=-(B1.Xt1)*Kd*Ky/Abs[-Kg-Ky*Kd*Udt2];

$$N2 = \begin{pmatrix} 1 & D2[[1]] \\ 0 & 1+D2[[2]] \end{pmatrix};$$

Hs=At2.N2.At1;

Sei=Eigenvalues[H1]

Abs[Sei]

In this cell, **Udt2** corresponds to the derivative $u_i(mT + \tau)$, **Hs** denotes the matrix H, and **Sei** denotes the eigenvalues of the matrix H. Mathematica outputs the following values:

$$\{0.996663, 0.996663\}$$

Increasing the gain **Ky** further to 1.5 yields

$$\{1.01795,\ 1.01795\}$$

In this case the considered system becomes unstable.

In the next cell we determine the steady-state values using the averaged method

Au:=A1*Du+A2*(1-Du);

Xu:=-Inverse[Au].Ev;

NsU=NSolve[Ky*Uref/Ug-Ky*Kd*Part[Xu,2]/Ug-Du== 0, Du]

Dd=Du/.NsU[[3]]

Ad=Au/.{Du->Dd};

Xd=-(Inverse[Ad].Ev)/.{Du->Dd}

In this cell, **Au** denotes $A(D)$, and **Xu** denotes $\hat{X}$. These variables are used for solving the nonlinear algebraic equation (3.37). Variables **Dd**, **Ad**, and **Xd** denote D, $A(D)$, and $\hat{X}$ for the steady-state process.

Since the calculation of **NsU** gives

$$\{\{Du\text{->}1.23341\},\quad \{Du\text{->}1.00016\},\quad \{Du\text{->}0.726798\}\},$$

we choose the third value (because its value is less than 1 and greater than 0) and substitute it in **Ad** and **Xd.** Then we calculate the eigenvalues of the matrix F (3.40)

As1=Outer[Times,((A1-A2).Xd),Ii]*Kd*Ky/Ug;

Fs=Ad-As1;

Eigenvalues[Fs]

which yields

$$\{-3689.55+58724.3\ i,\ -3689.55-58724.3\ i\}$$

Note that the **Outer[ ]** function with the option **Times** gives the outer product of the arguments **(A1-A2).Xd**, and **Ii**.

Since eigenvalues have negative real parts, the system according to the state-space-averaged method should be stable, but this contradicts the result obtained by the exact method. This situation is governed by the simplification introduced by the state-space-averaged method that cannot take into account pulsation of voltages and currents in the circuits of the converter.

Let us consider the behavior of the solutions and the stability conditions as $T \to 0$. The steady-state value (3.36) and the matrix F in (3.40) do not depend on the period T. The steady-state value of (3.33) as $T \to 0$ yields

$$\hat{X} = \lim_{T \to 0} \hat{X}(0) = -[A_2(1-D) + A_2 D]^{-1} \cdot [B_2(1-D) + B_1 D] = -A(D)^{-1} B.$$

With this we take into account the expressions $\frac{\tau}{T} = D$ and $\frac{T-\tau}{T} = 1 - D$. Calculating the limit of (3.30) at $t = mT + \tau$, we obtain

$$\hat{X} = \lim_{T \to 0} X(mT + \tau).$$

Then Equation 3.34 takes the form

$$DU_{ag} = k[u_{ref} - k_r \hat{\tilde{u}}].$$

The obtained expressions correspond to expressions (3.36) and (3.37).

The stability condition of the linearized system and the system with averaged-state variables are obtained for the difference (3.24) and differential equations (3.39). In order to compare these equations, we write the solution to Equation 3.39 on the interval equal to the period T:

$$\bar{X}_\xi((m+1)T) = e^{FT} \bar{X}_\xi(mT).$$

The form of this solution suggests an approach for comparing stability conditions.

Let us find a matrix H that satisfies the equality

$$e^{HT} = e^{A_2(T-\tau)} N(\tau) e^{A_1 \tau}$$

as $T \to 0$. We write a matrix exponential as a series and limit it with a few terms:

$$I + HT + \frac{1}{2} H^2 T^2 ... = [I + A_2(1-D)T + ..][I + \tilde{N}][I + A_1 DT + ...] \qquad (3.41)$$

where $\tilde{N} = D(\tau)G$.

In the expression $u_{tcom}(\tau) = |-kk_r u_t(\tau) - \frac{U_g}{T}|$, the derivative $u_t(\tau) \to 0$, and $\frac{U_g}{T} \to \infty$ as $T \to 0$. Therefore, one can consider that

$$u_{tcom}(t) \to \frac{U_g}{T}$$

as $T \to 0$, and then,

$$\tilde{D} = kk_r \frac{A_d \hat{\bar{X}} + B_d}{U_g} T = \tilde{D}_0 T,$$

where $\tilde{D}_0 = k_e(A_d\hat{X} + B_d)$. Multiplying the expressions on the right-hand side of (3.41), we obtain

$$I + HT + \frac{1}{2}H^2T^2... = I + [A_2(1-D) + A_1D + \tilde{N}_0]T + MT^2 + ... \qquad (3.42)$$

where $\tilde{N}_0 = \tilde{D}_0G$; and M is a matrix.

The expression (3.42) can be rewritten as follows:

$$H + \frac{1}{2}H^2T... = A_2(1-D) + A_1D + \tilde{N}_0 + MT + ...$$

Therefore,

$$H = A_2(1-D) + A_1D + \tilde{N}_0.$$

For a sufficiently small T the vector,

$$\tilde{D}_0 = -k_e \begin{vmatrix} \dfrac{\hat{u}}{L} \\[2mm] -\dfrac{\hat{i}}{C} \end{vmatrix}$$

coincides with the vector $A_d\hat{X}$. Therefore, the matrix H coincides with the matrix F. It has been shown that the state-space-averaging method gives the same results as the exact method for a sufficiently small period T.

Appropriate application of the described methods depends to a large extent on the examined question. Simplicity in use is a great advantage of the state-space-averaging method. Its disadvantage is the absence of the accuracy estimation.

3.4 Steady-State and Chaotic Processes in Closed-Loop Systems with PWM

For the description of the behavior of processes in nonlinear systems, a notion of attractor has been introduced, which generalize notions of the equilibrium position, limit cycle, and quasi-periodical process. The position of an equilibrium point, and periodic and quasi-periodical processes, exists in a system when a stability condition is executed.

In a system, chaotic processes could exist, characterized by an irregularity of motions. Such motions are connected with the instability of a system, but, at the same time, trajectories do not leave a bounded area in the state space. The domain of attraction in which chaotic motions exist is called a strange attractor.

An important virtue of a nonlinear system is the virtue of dissipativeness. For the system in which processes are described by the stationary differential equation

$$\frac{dX}{dt} = F(X) \tag{3.43}$$

where $X = (x_1, x_2, ..., x_n)$ is the vector, and $F(X) = (F_1(X), F_2(X), ..., F_n(X))$ is the nonlinear vector, the virtue of dissipativeness can be determined by the divergence theorem

$$\frac{1}{V}\frac{dV}{dt} = \sum_i \frac{\partial F_i}{\partial x_i}, \tag{3.44}$$

where V is the region in the state space.
  If the inequality

$$\sum_i \frac{\partial F_i}{\partial x_i} < 0 \tag{3.45}$$

holds, then the system will be dissipative. The realization of this condition shows a possibility of the existence of a strange attractor.
  For a second-order system, the condition of dissipativeness is based on Green's theorem, which permits writing the equation of changing of area S:

$$\frac{dS}{dt} = \iint \left(\frac{\partial Q}{\partial x} + \frac{\partial P}{\partial y} \right) dxdy,$$

where $Q = F_1(x, y)$, $P = F_2(x, y)$.
  The behavior of the system with PWM described by Equation 3.43 is only valid for separated intervals of constancy structure. A general set of equations is nonlinear and nonstationary. In that case, it is necessary to obtain the nonlinear difference equation in which state variables are connected with initial values. These values are obtained for the time moments at which the structure of a system is changed. For the i-th interval of constancy of structure $(t_{i-1} \leq t < t_i, i = 1, 2, ..., N)$, the differential equation has the form

$$\frac{dX(t)}{dt} = A_i X(t) + B_i \tag{3.46}$$

where A_i, B_i is the matrix and vector.
  Solving Equation 3.46 for all intervals and linking the initial $X(t_{i-1})$ and terminal $X(t_i)$ values for all intervals, we determine the difference equation

$$X((m+1)T) = G(X(mT)). \tag{3.47}$$

This equation is stationary and nonlinear. For this equation, the dissipative criterion (3.45) takes the form

$$\left| \det\left(\frac{\partial G_i}{\partial x_j} \right) \right| < 1. \tag{3.48}$$

In the expression (3.48), the matrix $\frac{\partial G_i}{\partial x_j}$ actually defines the linear approximation. In what follows, in order to determine the matrix $\frac{\partial G_i}{\partial x_j}$, we will apply the method of linearization described earlier.

In order that system motions correspond to a strange attractor, it is necessary for the existence of the following conditions: a sensitive dependence of phase space trajectory on initial conditions, a constraint of the area occupied by trajectories, and a contraction of the area. The first condition is connected with the stability condition. The second condition is satisfied for closed-loop systems. The third condition is characterized by the dissipativeness of the system.

The determination of the conditions of existence for a strange attractor will be realized with respect to an examined process whose equation has been linearized. As result of the linearization of Equation 3.47 for n periods, we obtain

$$X((m+n)T) = \prod_{k=1}^{n} H_k X(mT). \tag{3.49}$$

The stability and dissipativeness are determined on the basis of the analysis of the eigenvalues of matrix multiplication:

$$H = \prod_{k=1}^{n} H_k. \tag{3.50}$$

In this case, the conditions of existence of the strange attractor are formulated as follows:

The eigenvalues of the matrix H must be greater than unity.

The absolute value of the determinant of the matrix must be less then unity, that is,

$$|\det H| < 1. \tag{3.51}$$

Let us consider the Buck converter with PWM (Figure 3.15) (Zhuykov and Korotyeyev, 2000).

The electromagnetic processes in the converter circuits are described by the matrix differential equation

$$\frac{dX(t)}{dt} = AX(t) + B, \tag{3.52}$$

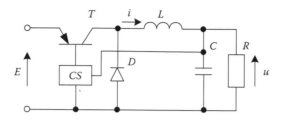

FIGURE 3.15
Buck converter.

where

$$X(t) = \begin{vmatrix} i \\ u \end{vmatrix}$$

is the vector of state variables;

$$A = \begin{vmatrix} -\dfrac{r}{L} & -\dfrac{1}{L} \\ \dfrac{1}{C} & -\dfrac{1}{RC} \end{vmatrix} ; \quad B = \begin{vmatrix} \dfrac{E\gamma}{L} \\ 0 \end{vmatrix}.$$

For steady-state stability analysis, the matrix H, for process with the period nT, takes the form

$$H_k = e^{A(T-\tau_k)}N_2(\tau_k)e^{A\tau_k}, \tag{3.53}$$

where

$$N_2(\tau_k) = \begin{vmatrix} 1 & d_2^1(\tau_k) \\ 0 & 1+d_2^2(\tau_k) \end{vmatrix} ; \quad D_2(\tau_k) = -kk_r \frac{B_\gamma(\gamma)}{|u_{tcom}(mT+\tau_k)|}$$

$$= -\frac{kk_r}{k_2}\begin{vmatrix} \dfrac{E}{L} \\ 0 \end{vmatrix} ; \quad D_2(\tau_m) = \begin{vmatrix} d_2^1(\tau_k) \\ d_2^2(\tau_k) \end{vmatrix}.$$

Using the condition (3.51), and with $\det N_2(\tau_m) = 1$, we obtain

$$|\det H| = \prod_{k=1}^{n} e^{A_1(T-\tau_k)}e^{A_1\tau_k}.$$

It follows that the presented system is always dissipative.

Let us calculate the processes in the converter circuit for the following parameter values: $E = 120$ V; $C = 12.5$ µF; $L = 8$ mH; $R = 8\ \Omega$; $r = 0.4\ \Omega$; $T = 0.3$ ms; $u_{ref} = 10$ V; $U_g = 4$ V; $k_r = 0.125$, $k = 4.4$. In the cell, the parameter values, and expressions for the matrix and vectors are defined:

$$R1=0.4;$$

$$L1=8.0*10^{\wedge}(-3);$$

$$C1=12.5*10^{\wedge}(-6);$$

$$Rn=8;$$

$$E1=120;$$

$$A1 = \begin{pmatrix} -R1/L1 & -1/L1 \\ 1/C1 & -1/(Rn*C1) \end{pmatrix};$$

$$B1 = \begin{pmatrix} E1/L1 \\ 0 \end{pmatrix};$$

$$Ug=4.0;$$

$$Uref=10.0;$$

$$T=0.3*10^{\wedge}(-3);$$

$$Ky=4.4;$$

$$Kr=0.125;$$

$$X0 = \begin{pmatrix} 0 \\ 0 \end{pmatrix};$$

$$Kg=Ug/T;$$

$$Iu=\{0, 1\};$$

$$Ii=\{1, 0\};$$

In this cell, **R1** denotes r; **Rn** denotes R; Ky denotes k; **Ug** defines the voltage amplitude of the independent generator; the vectors **Ii** and **Iu** select the first and second components; and the vector **B1** defines the value of the vector B for $\gamma = 1$. In the next cell

$$At1:=MatrixExp[A1*t1];$$

$$At2:=MatrixExp[A1*t2];$$

$$A1inv=Inverse[A1];$$

$$I2=IdentityMatrix[2];$$

Xn[1]=X0;

n=1;

Xt1:=At1.Xn[n]+A1inv.(At1-I2).B1;

Uc:=-Kg*t1+Ky*(Uref-Kr*Part[Xt1,2]);

the functions for the matrix exponentials, the vector of state variables for the switching moment, and the voltage on the input of the comparator are defined. For the calculation of switching moments and the transient process in the cell

Nmax=400;

Nmin=380;

Np = 8;

delta=1*10^(-8);

For[n=1,n<=Nmax,n++,{aa=0,bb=T,Xn[n]=X0,While[(bb-aa)/T>delta,

{t1=(aa+bb)/2,If[Part[Uc,1]<0,bb=t1,aa=t1]}],tn[n]=t1,Xnt1[n]=Xt1,t2=T-t1,

Xn[n+1]=At2.Xnt1[n],X0=Xn[n+1]}];

the dichotomy method is employed. The variable **Nmax** defines the maximum number of periods for the calculated process; **Nmin** defines the number of periods used to plot the phase-plane portrait; and **Np** defines the number of periods used to plot the transient process.

In the next cell, the functions are used to plot the phase-plane portrait, generator and control voltages are defined:

Zif[t_, n_]:=If[(t<(n-1)*T+tn[n])&&(t>=(n-1)*T),MatrixExp[A1*

(t−(n-1)*T)].Xn[n]+A1inv.(MatrixExp[A1*(t-(n-1)*T)]-I2).B1,

If[(t<n*T)&&(t>=(n-1)*T+tn[n]),

MatrixExp[A1*(t-(n-1)*T-tn[n])].Xnt1[n],0]];

Y3[t_]:=Sum[Zif[t,n],{n,1,Nmax}];

Zg[t_,n_]:=If[(t<=n*T)&&(t>(n-1)*T),Kg*(t-(n-1)*T),0];

Yg[t_]:=Sum[Zg[t,n],{n,1,Nmax}];

Zoc[t_, n_]:=If[(t<(n-1)*T+tn[n])&&(t>=(n-1)*T),Ky*(Uref-Kd*

Part[(MatrixExp[A1*(t-(n-1)*T)].Xn[n]+A1inv.(MatrixExp[A1*

(t−(n-1)*T)]-I2).B1),2]),If[(t<n*T)&&(t>=(n-1)*T+tn[n]),Ky*(Uref -

Kd*Part[(MatrixExp[A1*(t-(n-1)*T-tn[n])].Xnt1[n]),2]),0]];

Yoc[t_]:=Sum[Zoc[t,n],{n,1,Nmax}];

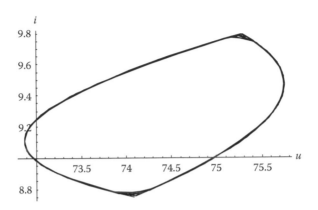

FIGURE3.16
Phase-plane portrait for $k = 4.4$ (i in amperes, u in volts).

In Figure 3.16 the phase-plane portrait is presented, and in Figure 3.17 the generator and control voltages. For plotting the process we use the functions

Yt=Plot[Yg[t],{t,(Nmax-Np8)*T,Nmax*T},

AxesLabel->{"t","u"},AxesOrigin->{(Nmax-Np)*T,0},

DisplayFunction->Identity];

Poc=Plot[Yoc[t],{t,(Nmax-Np)*T,Nmax*T},AxesLabel->{"t","u"},

AxesOrigin->{(Nmax-Np)*T,0},DisplayFunction->Identity];

Show[{Poc,Yt},DisplayFunction->$DisplayFunction];

ParametricPlot[{Part[Iu.Y3[t],1],

Part[Ii.Y3[t],1]},{t,Nmin*T,Nmax*T},AxesLabel->{"u","i"},

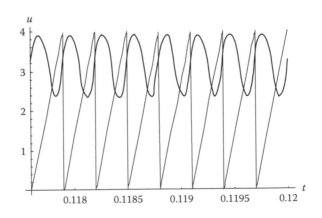

FIGURE 3.17
Generator and control voltages for $k = 4.4$ (u in volts, time t in seconds).

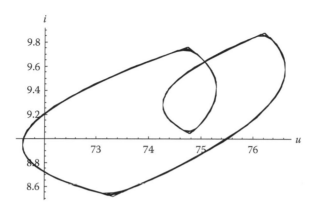

FIGURE3.18
Phase-plane portrait for $k = 4.6$ (i in amperes, u in volts).

Data for plotting the generator and control voltages are assigned to two functions, **Yt** and **Poc**. Both voltages are plotted simultaneously with the help of the **Show[ ]** function.

By increasing the gain, a bifurcation takes place and, in the system, a process is formed with the period $2*T$. In Figures 3.18 and 3.19, the phase-plane portrait and the voltages for $k = 4.6$ are presented.

When $k \approx 9.4$, a new bifurcation takes place in the system, and a process with the period $4*T$ is formed. In Figures 3.20 and 3.21, the phase-plane portrait and the voltages for $k = 9.6$ are presented.

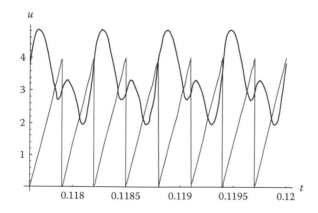

FIGURE 3.19
Generator and control voltages for $k = 4.6$ (u in volts, time t in seconds).

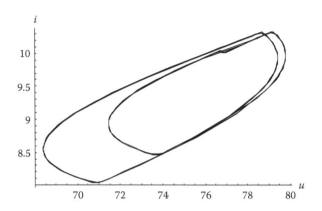

FIGURE 3.20
Phase-plane portrait for $k = 9.6$ (i in amperes, u in volts).

The determination of process stability is accomplished by the computation of the eigenvalues of the matrix H_k for intervals with the duration T.

$$N2n := \begin{pmatrix} 1.0 & -d12*E1/L1 \\ 0 & 1.0 \end{pmatrix};$$

For[n=1,n<=Nmax,n++,{t1=tn[n],t2=T-t1,

Udt=Part[Iu.(A1.Xnt1[n]+B1),1],

d12=Kd*Ky/Abs[-Kg-Kd*Ky*Udt],If[tn[n]>0.99*T,f12p=0,0],

Hm[n]=At2.N2m.At1}];

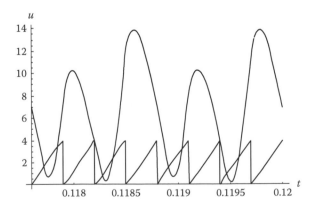

FIGURE 3.21
Generator and control voltages for $k = 9.6$ (u in volts, time t in seconds).

It should be stressed that, in the program, the limitation of maximum pulse duration **tn[n] > 0.99*T** is introduced. This is connected with the fact that impulse duration theoretically can be greater than the period.

In the next cell, the eigenvalues for the end of computed intervals are outputted:

Print["Nmax ",Eigenvalues[Hm[Nmax]]];

Print["Nmax-1 ",Eigenvalues[Hm[Nmax-1]]];

Print["Nmax-2 ",Eigenvalues[Hm[Nmax-2]]];

Print["Nmax-3 ",Eigenvalues[Hm[Nmax-3]]];

Print["Nmax*(Nmax-1) ",Eigenvalues[Hm[Nmax].Hm[Nmax-1]]];

Print["Nmax*...*(Nmax-3) ",Eigenvalues[Hm[Nmax].Hm[Nmax-1].

Hm[Nmax-2].Hm[Nmax-3]]];

The values of gains and eigenvalues are presented in Table 3.1. In the second column of the table, the eigenvalues for the interval of duration T are presented. In the forth column, the eigenvalues are presented for an interval equal to the period of steady-state process. For all presented processes, det $|H_m| = 0.049$.

For the gain k ≈ 27.2, the absolute values of eigenvalues determined for the arbitrary-interval aliquot to the period become greater than unity. There the system remains dissipative since the condition (3.51) holds for any number of intervals. Thus, the system is unstable and dissipative at the same time. In this case, it can be argued that the examined process corresponds to the strange attractor (Figure 3.22). The calculation is made for an interval equal to 800T, when the plotting is realized for 700$T \le t \le$ 800T. It should be noted that, by increasing the time interval used for plotting, the phase trajectories fill the bounded area.

TABLE 3.1

The Values of Eigenvalues for Different Gains and Periods of Steady-State Process

Gain, k	Eigenvalues for the Interval T	Period of Steady-State Process	Eigenvalues for the Period
4.4	−0.98, −0.05	T	−0.98, −0.05
4.6	−1.105, −0.044	2T	0.9136, 0.0026
	−0.917, −0.053		
9.6	0.7, 0.07	4T	0.41, 0.000014
	−1.34, −0.037		
	−1.26, −0.039		
	−1.15, −0.043		

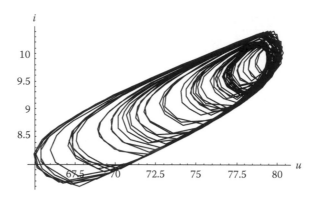

FIGURE 3.22
Phase-plane portrait of the strange attractor for $k = 28$ (i in amperes, u in volts).

3.5 Identification of Chaotic Processes

Nonregular motions in a system can be connected with various phenomena. In the first place, in a system, there could be quasi-periodic oscillations with a few incommensurable frequencies. In the second place, a strange attractor could arise in a system. There is also the possibility of error connected with the fact that an investigated interval is chosen inside a transient process. For the identification of processes, the following operations are used: Poincare section, computation of the attractor dimension, Lyapunov exponent, and the correlation function (Strzelecky et al., 2001).

The map P connecting the coordinates of points in which the trajectory of the motion of a system intersects a given surface

$$a_{m+1} = P(a_m)$$

is called a Poincare section. With the help of the Poincare section, the transition from a system with continuous time to a system with discrete time is achieved.

Another method of finding a Poincare section is based on the solution to Equations 3.1 and 3.5 at defined moments $mT + t_m$ (T is the interval of sampling, and t_m is the time moment inside the interval T). In this case the equation connecting the sampled point has the form

$$X_{m+1} = P(X_m), \tag{3.54}$$

where X_m is the value of the vector X at the time $mT + t_m$. For the nonstationary system (3.5) with a periodic forcing function, it is expedient to choose the step T equal to the period of this function. In this case, Equation 3.54 can be written in the form (3.47).

A Poincare section of a periodical process has only one point. If in a system subharmonic oscillations with period 2*T* arise, the Poincare section contains two points, whereas, if quasi-periodical oscillations arise, the Poincare section contains a closed graph. Poincare sections of strange attractors represent point sets that form groups in some way.

The plotting of a Poincare section for voltage across the capacity for time moments *t* = *mT* on the basis of data obtained earlier is done as follows:

Nu=400;

PuancareUn=Table[{Part[Iu.Xn[n],1],Part[Iu.Xn[n+1],1]},{n,Nu,Nmax-1}];

ListPlot[PuancareUn,AxesLabel->{"Un","Un+1"}];

The variable **Nu** defines the initial value for the output of points. During the calculation, the following interval value **Nmax = 800** was used. The Poincare section is presented in Figure 3.23. When increasing the time interval used for output, the points are located practically on the same curve. Therefore, the process under consideration is chaotic, and the attractor is strange.

An attractor dimension characterizes the number of degrees of freedom of points corresponding to this attractor. For a subset in the phase space occupied by an attractor, the attractor dimension is defined by the expression

$$d_0 = \lim_{\varepsilon \to 0} \frac{\ln N(\varepsilon)}{\ln\left(\frac{1}{\varepsilon}\right)},$$

where $N(\varepsilon)$ is the minimum number of cubes with cube size ε that are necessary for covering a subset. This expression is the definition of the Hausdorff dimension.

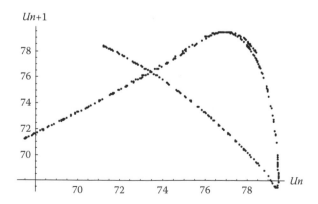

FIGURE 3.23
Poincare section of the voltage across capacity for $k = 28$ (U_{n+1} and U_n in volts).

For a point attractor, the number $N(\varepsilon) = 1$ and does not depend on length ε, and therefore, $d_0 = 0$. When the attractor is a closed cycle, then $N(\varepsilon) \sim \varepsilon$, and $d_0 = 1$ (taken into account are the squares through which the cycle curve is run). In the case when the domain occupied by the attractor is a surface, then $N(\varepsilon) \sim \varepsilon^2$ (squares covering the inner surface are taken into account), and therefore, $d_0 = 2$. The strange attractor does not have integral dimension; the Lorenz attractor has the dimension $d_0 = 2.06$, the Henon attractor has $d_0 = 1.25$, and the logistic attractor has $d_0 = 0.543$.

In calculating a process dimension, it is convenient to use the definition of a dimension on the basis of the correlation function

$$C(r) = \frac{1}{N_M^2} \sum_i \sum_j H(r - \|x_i - x_j\|), \qquad (3.55)$$

where N_M is the number of points; r is the radius of the circle in the point x_i; $\|\dots\|$ is the distance between points x_i and x_j; and H is the Heaviside step function. The correlation dimension d_2 is defined by the expression

$$d_2 = \lim_{r \to 0} \frac{\ln C(r)}{\ln r}. \qquad (3.56)$$

Let us consider a dimension change for processes running in the circuit of the Buck converter (Section 3.4). The calculation of the correlation function (3.55) is executed as follows:

Clear[k,n,m];

rk=0.2;

Nk=20;

Nmin=Nmax-100;

Norma:=Sqrt[(Iu.Xn[n]-Iu.Xn[m])^2+(Ii.Xn[n]-Ii.Xn[m])^2];

For[k=1,k≤Nk,k++,{Cr[k]=0;rr=rk*k;

For[m=Nmin,m≤Nmax,m++,For[n=Nmin,n≤Nmax,n++,

If[(rr≥Norma[[1]]),Cr[k]=Cr[k]+1,1]]]}];

In this cell the variable **rk** defines the minimum circle size, and **Nk** the number of points or radius **rr**. The argument of Heaviside step function is determined with the help of the **Norma** function.

In view of the finite number of computed intervals, the results of the calculation of the dimension by (3.56) are inexact. The presentation of the results of the calculation of the dimension seems to be well expedited by the graph of the $\ln C(r) = f(\ln r)$ function.

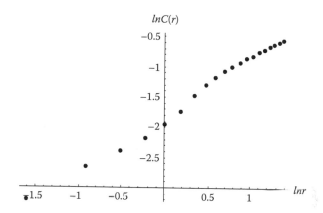

FIGURE 3.24
Dependence ln $C(r) = f(\ln r)$ for $k = 28$.

In the next cell the values of logarithms of correlation function and radius are calculated by forming the table

TabP=Table[{Log[rk*i],Log[Cr[i]/((Nmax-Nmin)^2)]},{i,Nk}];

ListPlot[TabP,Prolog->AbsolutePointSize[4]]

The size of the plotting points is defined by the **Prolog->AbsolutePointSize[4]** option. The graph of the ln $C(r) = f(\ln r)$ function is presented in Figure 3.24. For $k = 9.6$ (subharmonic oscillation with the period $4T$), the dependence ln $C(r) = f(\ln r)$ is shown in Figure 3.25. Comparing Figures 3.24 and 3.25, one can see the qualitative change of the dependence ln $C(r) = f(\ln r)$.

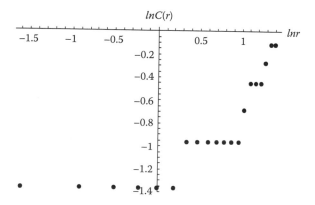

FIGURE 3.25
Dependence ln $C(r) = f(\ln r)$ for $k = 9.6$.

The correlation function for the continuous process is defined by the expression

$$K(\tau) = \lim_{A\to\infty} \frac{1}{A} \int_0^A \hat{x}(t)\hat{x}(t+\tau)\,dt,$$

where $\hat{x}(t) = x(t) - \lim_{A\to\infty} \frac{1}{A}\int_0^A x(t)\,dt$; for periodical processes, the function is periodical: $K(T+\tau) = K(\tau)$. For chaotic processes, the correlation function behaves in the following way: $\lim_{\tau\to\infty} K(\tau) = 0$.

For processes described by the difference equation (3.47), the correlation function is defined as follows:

$$K(m) = \lim_{N_M\to\infty} \frac{1}{N_M} \sum_{i=0}^{N_M-1} \hat{x}_{i+m}\hat{x}_i,$$

where

$$\hat{x}_i = x_i - \bar{x}; \quad \bar{x} = \lim_{N_M\to\infty} \frac{1}{N_M} \sum_{i=0}^{N_M-1} x_i.$$

We perform a calculation of the correlation function on the finite interval **Nt**:

$$K(m) = \frac{1}{N_M - N_k} \sum_{i=1}^{N_M-N_k} \hat{x}_{i+m}\hat{x}_i, \tag{3.57}$$

where N_k is the number of points to be calculated in the correlation function $(1 \le m \le N_k)$. The limitation of the interval is related to the need not to exceed the number of interval **Nmax**.

Nt=200;

Xav=Sum[Iu.Xn[k],{k,1,Nmax}]/Nmax;

For[n=1,n≤Nt,n++,Kor[n]=Sum[(Iu.Xn[i]-Xav)*(Iu.Xn[i+n]-Xav),

{i,1,Nmax-Nt}]/(Nmax-Nt)];

The variable **Xav** corresponds to the average value $\bar{x}$, and **Kor** is the correlation function. The plotting of the correlation function is done as follows:

TabKor=Table[{i,Part[Kor[i],1]},{i,Nt}];

ListPlot[TabKor,Prolog-> AbsolutePointSize[2]]

If a process is not chaotic, the correlation function represents a regular function. Figure 3.26 shows the correlation function of the process for $k = 9.6$ and,

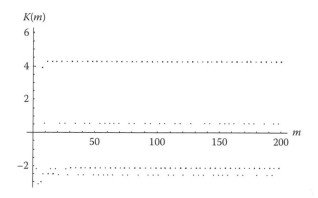

FIGURE 3.26
Correlation function of the process for $k = 9.6$.

in Figure 3.27, for $k = 28$. The randomness of the correlation function allows one to come to the conclusion that a strange attractor is present in the system. The irregular behavior of the correlation function allows one to distinguish chaotic processes from processes with an irrational relationship between frequencies, which at first glance could have a high resemblance to chaotic ones.

Analyses of convergence or divergence of processes is expediently carried out with the help of Lyapunov exponents. For the linear system

$$\frac{dX}{dt} = A(t)X,$$

the Lyapunov exponents are defined as

$$\alpha = \lim_{t \to \infty} \frac{1}{t} \ln \frac{\|X(t)\|}{\|X_0\|},$$

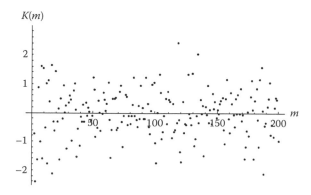

FIGURE 3.27
Correlation function of the process for $k = 28$.

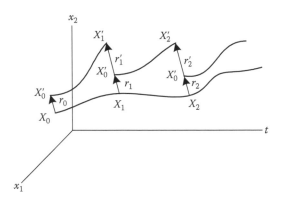

FIGURE 3.28
Processes of scaling at the calculation of the maximum Lyapunov exponent.

where $\|\cdots\|$ is the norm; and X_0 is the initial value of the vector $X(t)$. The number of Lyapunov exponents corresponds to the phase space dimension. For the stationary system $A(t) = A = const$, the Lyapunov exponents $\alpha_i = \operatorname{Re}\lambda_i$ are determined by the eigenvalues λ_i of the matrix A.

The calculation of Lyapunov exponents for a nonlinear system is based on a numerical procedure of calculation of two processes starting close to each other. The calculation is carried out on intervals of duration such that the processes do not reach a value greater than the capacity of the computer used. At the beginning of every interval, the initial values of one of the processes are determined by scaling of calculated values obtained at the end of the previous interval (Figure 3.28).

The calculation of the two processes begins at initial values X_0 and X_0'. Further, the distance $r_0 = \sqrt{(x_{10} - x_{10}')^2 + (x_{20} - x_{20}')^2}$ between the initial values is chosen sufficiently small. Calculation is continued for the moment t_1. This moment is determined by an experimental method and is connected with the rate of process divergence. Calculation of the next interval $t_1 \le t \le 2t_1$ for one of the processes is realized for initial values

$$X_0' = X_1 + r_0(X_1' - X_1)/r_1'.$$

With respect to such scaling $r_0 = r_1 = r_2 = .. = r_N$, the maximum value of the Lyapunov exponent is determined by calculating the expression

$$\alpha_M = \lim_{N \to \infty} \frac{1}{N t_1} \sum_{i=1}^{N} \ln \frac{r_i'}{r_0}, \tag{3.58}$$

where r_i' is the distance between processes at the end of i-th interval.

The calculation of the maximum value of Lyapunov exponents for various values of gain is realized as follows:

$$ui0=0.00001;$$

$$i0=ui0;$$

$$u0=ui0;$$

$$NL=20;$$

$$KL=Floor[Nmax/NL];$$

$$r0=Sqrt[i0^2+u0^2];$$

$$kL=0;$$

$$MLyap=61;$$

$$For[m=1,m<=MLyap,m++,\{Ky=4+0.4*(m-1),kL=0, X2 = \begin{pmatrix} i0 \\ u0 \end{pmatrix}, X1 = \begin{pmatrix} 0 \\ 0 \end{pmatrix},$$

$$For[n=1,n<=Nmax,n++,\{For[k=1,k<=2,k++,\{aa=0,bb=T,If[k==1,Xn[n]=X1,$$
$$Xn[n]=X2],$$

$$While[(bb-aa)/T>delta,\{t1=(aa+bb)/2,$$

$$If[Part[Uc,1]<0,bb=t1,aa=t1]\}],tn[n]=t1,Xnt1[n]=Xt1,t2=T-t1,$$

$$Xn[n+1]=At2.Xnt1[n],If[k==1,X1=Xn[n+1],X2=Xn[n+1]]\}],$$

$$If[Mod[n,NL]==0,\{kL=kL+1,ri=Part[Sqrt[(Ii.(X2-X1))^2+(Iu.(X2-X1))^2],1],$$

$$SSn[kL]=Log[ri/r0],X22=X2,X2=X1+r0*(X22-X1)/ri\},1]\}],$$

$$LyE[m]=Sum[SSn[nn],\{nn,1,KL\}]/(T*Nmax) \}];$$

where the variable **ui0** defines the initial values of the current and voltage (chosen in such a way that the distance **r0** between processes would be sufficiently small); **NL** defines the number of intervals of duration T in such a way that $NL \cdot T = t_1$; **KL** corresponds to the number of intervals N; **kL** corresponds to variable i in (3.58); **MLyap** defines the number of calculating values for gain **Ky**; and **Ky=4+0.4*(m-1)** defines the range of change of the gain.

For outputting the values of the Lyapunov exponent, a table is formed. To plot the graph of the function, the option **PlotJoined->True** is used, which allows the combination of calculated points by straight-line segments.

$$TabLyap=Table[\{4+0.4*(i-1),LyE[i]\},\{i,MLyap\}];$$

$$ListPlot[TabLyap,Prolog->AbsolutePointSize[2],$$

$$PlotJoined->True,AxesLabel->\{"Ky","Lyapunov Exponent"\}];$$

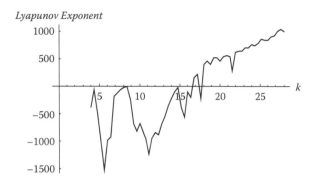

FIGURE 3.29
Dependence of the maximum value of Lyapunov exponents versus gain.

The dependence of the maximum value of Lyapunov exponents is presented in Figure 3.29. As one sees from the figure, a chaotic process can emerge in the system for $k > 16.7$.

3.6 Calculation of Processes in Relay Systems

In relay systems, the forming of alternating voltage on a load is based on the tracing of a given sinusoidal signal u_g (Figure 3.30).

In such a system the power supply of a converter is provided by the DC voltage E. The control of the converter is handled in such a way that, on its output, rectangular impulses are formed whose frequencies and duty factors are determined by a dead band of a relay element. A sinusoidal voltage generation on the load L is made by the output of filter F. In Figure 3.31 the block diagram of the relay system as relay controller is shown. Assume that the controller is proportional, and the filter and the load (Figure 3.32) are described by a transfer function $W(p)$ of the second order.

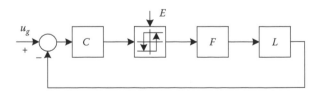

FIGURE 3.30
Block diagram of a relay system.

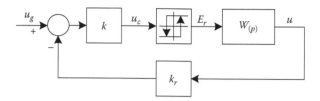

FIGURE 3.31
Block diagram of the relay system as relay controller.

The dynamic characteristics of a closed system are determined by the parameters of the relay characteristics, the filter and the load, and behavior of the controller. We will analyze the conditions for the onset of self-oscillations with the help of harmonic linearization (Korotyeyev, 2003a). The electromagnetic processes in a closed-loop system are described by the equations

$$\frac{di}{dt} = -\frac{r}{L}i - \frac{1}{L}u + \frac{E_r}{L};$$

$$\frac{du}{dt} = \frac{1}{C}i - \frac{1}{RC}u; \tag{3.59}$$

$$u_c = k(u_g - k_r u);$$

$$E_r = f(u_c),$$

where E_r is the voltage on the output of the relay element (Figure 3.33); $u_g = U_g \sin(\omega t + \varphi)$; k is the gain of the proportional regulator; and k_r is the output voltage ratio.

The method of harmonic linearization is used for the analysis of processes in the closed-loop system and is based on the investigation of the first

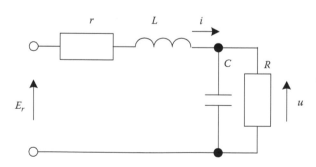

FIGURE 3.32
Circuit of the filter and load.

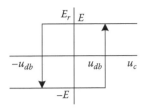

FIGURE 3.33
Characteristic of the relay element.

harmonic passing through a system. According to this method, we write the expression of an output voltage of a nonlinear element in the form

$$J(U_g) = q(U_g) + jq'(U_g), \tag{3.60}$$

where $q(U_g) = \frac{4E\sqrt{U_g^2 - u_{db}^2}}{\pi U_g^2}$, $q'(U_g) = -\frac{4Eu_{db}}{\pi U_g^2}$ are the coefficients of harmonic linearization; and U_g is the amplitude of sinusoidal generator voltage.

The condition for the onset of oscillations in the closed-loop system in the absence of the external action u_g is determined by the expression

$$kk_r W(j\omega) = -\frac{1}{J(U_g)}, \tag{3.61}$$

where $W(p) = \frac{R}{LCRp^2 + (CRr + L)p + R + r}$ is the transfer function of the filter and load;

$$W(j\omega) = \frac{R(R + r - \omega^2 LCR)}{(R + r - \omega^2 LCR)^2 + \omega^2(CRr + L)^2} - \tag{3.62}$$

$$\frac{j\omega R(CRr + L)}{(R + r - \omega^2 LCR)^2 + \omega^2(CRr + L)^2}.$$

In Figure 3.34 the right and left parts of the expression (3.61) are presented.

On the line corresponding to the nonlinear function $-\frac{1}{J(U_g)}$, the use of the arrow shows the direction of the increasing value of this function with an increase in the amplitude U_g. Since the motion during the increase in value occurs from the area bounded by the amplitude-frequency characteristic, the cross point A is stable.

Equating the real and imaginary parts of Equation 3.61, we determine the frequency and amplitude of self-oscillations. Using expressions (3.61) and (3.62), we obtain the equation for frequency

$$\frac{\pi u_{db}}{4E}[(R + r - \omega^2 LCR)^2 + \omega^2(CRr + L)^2] - kk_r \omega R(CRr + L) = 0 \tag{3.63}$$

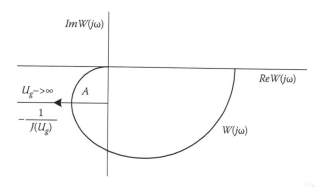

FIGURE 3.34
Amplitude-frequency characteristic of the filter and load, and linear characteristic of the non-linear element.

and the expression for the amplitude of self-oscillations

$$U_g = \frac{\sqrt{[4Ekk_r P(\omega)]^2 + (\pi u_{db})^2}}{\pi},$$ (3.64)

where $P(\omega)$ is the real part of the complex transfer function (3.62). At first, from Equation 3.63, the frequency of self-oscillation is determined, and then the $P(\omega)$ function and, thereafter, from expression (3.64), the amplitude of self-oscillation is calculated.

Let us determine the self-oscillation for the following parameter values:

r1=2.0;

Rn=12000.0;

L1=0.02;

C1=1.0*10^(-7);

E1=300.0;

Kr=1.0;

Ky=1.0;

Uref=200.0;

Ff=4636;

pg=2.0;

f0=0;

wf=2.0*Pi*Ff;

$$rL=(r1+Rn)/L1;$$

$$A1 = \begin{pmatrix} -R1/L1 & -1/L1 \\ 1/C1 & -1/(Rn*C1) \end{pmatrix};$$

$$EL:=E1/L1;$$

$$B1 := \begin{pmatrix} E1/L1 \\ 0 \end{pmatrix};$$

$$wkon=1/Sqrt[L1*C1];$$

$$I2=IdentityMatrix[2];$$

In this cell, **r1** denotes *r*, **Rn** denotes *R*, **Ky** denotes *k*, **Ug** denotes the amplitude of the sinusoidal signal u_g, **Ff** denotes the frequency of the sinusoidal signal u_g, and the vector **B1** defines the value of the vector *B* for the differential equation (3.59).

Using the expressions (3.63) and (3.64), we calculate the amplitude and frequency of self-oscillation:

$$descr=Expand[Pi*pg*((Rn+r1-\omega^2*L1*C1*Rn)^2$$
$$+ \omega^2*(C1*Rn*r1+L1)^2)/4/$$

$$Abs[E1]-Ky*Kr* \omega*Rn*(C1*Rn*r1+L1)];$$

$$desSol=Solve[descry==0, \omega]$$

$$\omega1=\omega/.desSol[[4]]$$

$$TpGarmLin=2*Pi/ \omega1$$

$$P\omega = Rn*(Rn+r1-\omega1^2*L1*C1*$$

$$Rn)/((Rn+r1- \omega1^2*L1*C1*Rn)^2+\omega1^2*(C1*Rn*r1+L1)^2)$$

$$Sqrt[(4*Abs[E1]*Ky*Kr*P\omega)^2+(Pi*pg)^2]/Pi$$

In this cell, **descr** defines Equation 3.63, **desSol** defines the solution to this equation, **TpGarmLin** defines the period of self-oscillation, and **Pω** denotes $P(\omega)$. The expression (3.64) is calculated in the last row of this cell. Mathematica outputs the roots of this expression as follows:

$$\{\{\omega->-27049.1-32492.4 \; i]\}, \{\omega->-27049.1 + 32492.4 \; i]\},$$

$$\{\omega->2723.39\}, \{\omega->51374.9\}\}$$

We calculate the amplitude of the frequency 51374.9 because, for that value, $P(\omega)$ is a negative −0.233605. Then the amplitude equals 89.2528.

Let us analyze the stability of oscillation with the help of linearization of the set (3.59). For the determination of stability conditions (Korotyeyev, 2003b), we will take into account that a change of an impulse front on the output of the relay element occurs at the beginning of every half-period of forced voltage, while initial and boundary conditions on that interval differ by the sign. A steady-state process is determined as a result of the solution to the two first equations of the set (3.59) on half of the period:

$$X(t) = e^{At}X(0) + A^{-1}(e^{At} - I)B,$$

where

$$A = \begin{vmatrix} -\dfrac{1}{r} & -\dfrac{1}{L} \\ \dfrac{1}{C} & -\dfrac{1}{RC} \end{vmatrix}; \quad B = \begin{vmatrix} \dfrac{E}{L} \\ 0 \end{vmatrix}; \quad X\left(\dfrac{T}{2}\right) = -X(0);$$

$$X(0) = \left(1 + e^{A\frac{T}{2}}\right)^{-1} A^{-1}\left(e^{A\frac{T}{2}} - 1\right)B.$$

The period of self-oscillation is determined by the use of the third equation of the set (3.59):

$$u_{db} = -kk_r u(0), \tag{3.65}$$

where $u(0)$ is the voltage across the capacitor in the steady state.

We realize, for the interval $mT \le t \le (m+1/2)T$, the stability analysis of the closed-loop system with an external sinusoidal voltage. Since there is one interval of constancy of structure, by linearization of the equation set (3.59), we obtain (3.17), in which t_μ is the time moment of switching of the relay element;

$$D_\mu = -kk_r \frac{2B}{|u_{tc}(t_\mu)|};$$

$$u_{tc} = \frac{du_c}{dt}\bigg|_{t=-0} = k(\omega U_g \cos\varphi - k_r u_t); \quad \text{and} \quad u_t = \frac{du}{dt}\bigg|_{t=-0} = \frac{1}{C}i(0) - \frac{1}{RC}u(0).$$

Solving Equation 3.17 for the interval $mT \le t \le (m+1/2)T$ yields

$$X((m+1/2)T) = e^{A\frac{T}{2}}NX(mT),$$

where $N = \begin{vmatrix} 1 & d_2^1 \\ 0 & 1 \end{vmatrix}$; $d_2^1 = -\frac{2Ekk_r}{|u_{tc}(0)|L}$. Then the matrix whose eigenvalues deter-

mine the stability of the linearized system is defined as

$$H = e^{A\frac{T}{2}} N \tag{3.66}$$

From the obtained expressions it follows that, for the next half-period of forced voltage, the stability condition remains invariable.

For stability calculation with the help of the linearization method, it is necessary first to find the solution to the nonlinear equation (5.65), which determines the period of self-oscillations:

E1=Abs[E1];

Clear[Ta];

At1:=MatrixExp[A1*Ta];

A1inv=Inverse[A1];

XT:=Inverse[(At1+I2)].A1inv.(At1-I2).B1;

Ua:=pg-Ky*Kr*Part[XT[[2]],1];

Tper=Ta/.FindRoot[Ua==0,{Ta,TpGarmLin}];

In this cell, **XT** denotes $X(0)$ for the steady-state process, **Ua** defines the equation (3.65). As a result, we obtain the period equal to **2*Tper**:

0.00012199 + 0.i

In the next cell, with the use of the matrix H (3.66), the stability conditions of self-oscillation are calculated:

Xt1:=At1.XT-A1inv.(At1-I2).B1;

Ta=Re[Tper];

Udt1=Part[Part[A1.Xt1,2],1]

f12s=-2*Abs[EL]*Ky*Kd/Abs[Ky*Kd*Udt1];

$$Q2s := \begin{pmatrix} 1.0 & f12s \\ 0 & 1 \end{pmatrix};$$

H1s:=At1.Q2s

Sei:=Eigenvalues[H1s]

Abs[Sei]

In this cell, **Xt1** denotes $X(T/2)$ the steady-state process, **Udt1** defines the derivative $u_{tc}(0)$, and **H1s** denotes the matrix H. The absolute values of the eigenvalues are {1., 0.944661}. The eigenvalue equal to unity indicates that, in the system, there arise oscillations whose phase depends on the initial conditions. Such oscillations are characterized by a stability (but not an asymptotical stability).

Let us consider the behavior of the system subject to the action of a sinusoidal voltage on its input. In this case, the system could have imposed oscillations from an external generator, natural oscillations could arise in the system or oscillations from an external source and from the system could exist at the same time. For verifying the existence of oscillations imposed by an external generator, it is necessary to study the stability of any obtained solutions. In the cell for the given half-period **tp**, the solution of the nonlinear differential equation obtained with the help of the periodicity condition is determined:

E1=Abs[E1];

Atp:=MatrixExp[A1*tp];

XTp:=Inverse[I2+Atp].A1inv.(Atp-I2).B1;

Xt1p:=Atp.XTp-A1inv.(Atp-I2).B1;

tn:=ArcSin[(Part[Part[Xt1p,2],1]-pg)/Uref];

Udt1:=Part[Part[A1.Xt1p,2],1];

ft:=tp*2;

wf:=2.0*Pi/ft;

f12s:=-2*Abs[EL]*Ky*Kd/Abs[Ky*Uref*wf*Cos[(Pi-tn)]-Ky*Kd*Udt1];

H1s:=Atp.Q2s;

In this cell, **XTp** and **Xt1p** denote $X(0)$ and $X(T/2)$ for the steady-state process, and **tn** defines the time at which the voltage goes to another part of the relay characteristic (Figure 3.33).

Now we use the functions defined in the previous cell to examine the behavior of the eigenvalues for different periods (in fact, for different half-periods):

t0=0.00002;

tnac=0.00004;

Nstep=1000;

For[n=1,n<=Nstep,n++,{tp=tnac+t0*n,

MaxExp[n]=Max[Abs[Eigenvalues[H1s]]]}];

TabMaxExp=Table[{tnac+t0*n,MaxExp[n]},{n,Nstep}];

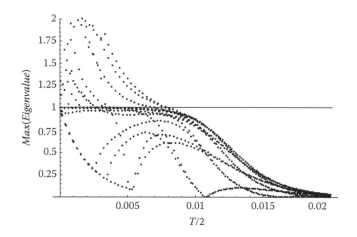

FIGURE 3.35
Dependence of the maximum absolute eigenvalue versus the half-period (T in seconds).

In this cell, **tnac, t0**, and **Nstep** define the first point, the step, and the number of steps, respectively. These variables are used for calculation eigenvalues for different periods. The calculated eigenvalues are placed in the table denoted **TabMaxExp.** The dependence of the maximum absolute eigenvalue versus the half-period is presented in Figure 3.35. For the plotting of the process we use the function

$$\textbf{ListPlot[TabMaxExp,AxesLabel->\{"t","Max|Eigenvalue|"\},}$$

$$\textbf{PlotRange->\{0,2.0\},GridLines->\{None,\{1.0,0.0\}\}];}$$

In this function, the option **GridLines->{None,{1.0,0.0}}** allows the generation of only one grid line passing through point one situated on the ordinate axis. For the obtained initial value **Xt1p**, the time moment is defined:

$$\textbf{tn:=ArcSin[(Part[Part[Xt1p,2],1]-pg)/Uref];}$$

which corresponds to the switching point of the relay element. For the half-period values 0.000002…0.002,

$$\textbf{t0=0.000002;}$$

$$\textbf{tnac=0.000004;}$$

$$\textbf{Nstep=1000;}$$

$$\textbf{For[n=1,n<=Nstep,n++,\{tp=tnac+t0*n,}$$

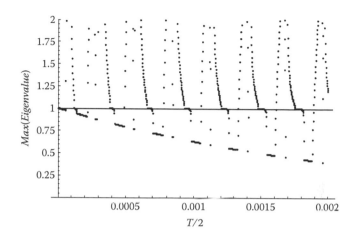

FIGURE 3.36
Dependence of the maximum absolute eigenvalue versus the half-period (*T* in seconds).

MaxExp[n]=Max[Abs[Eigenvalues[H1s]]]}];

TabMaxExp=Table[{tnac+t0*n,MaxExp[n]},{n,Nstep}];

ListPlot[TabMaxExp,AxesLabel->{"t","Max|Eigenvalue|"},

PlotRange->{0,2.0},GridLines->{None,{1.0,0.0}}];

the dependence of the maximum absolute eigenvalue versus the half-period is presented in Figure 3.36. For values less than unity in the system, the oscillations are formed with the frequency of an external generator. For values greater than unity in the system, subharmonic, quasi-periodical, or chaotic oscillations could be formed.

The calculation of transient processes is made using the expressions obtained from the solution to the differential equations. This calculation procedure allows appreciable reduction of calculation time in comparison with the procedure using expressions with a matrix exponential:

If[wf>wkon,Tp:=2*Pi/wkon,Tp:=2*Pi/wf];

If[Discrim>0,Tp:=2*Pi/wf,Tp:=2*Pi/wkon];

a11=-r1/L1;

a12=-1/L1;

a21=1/C1;

a22=-1/Rn/C1;

Discrim=(1/Rn/C1+r1/L1)*(1/Rn/C1+r1/L1)/4−(r1/Rn+1)/L1/C1;

If[Discrim>0,{p1=−(1/Rn/C1+r1/L1)/2+Sqrt[Discrim],

p2=−(1/Rn/C1+r1/L1)/2−Sqrt[Discrim]},{p1=−(1/Rn/C1+r1/L1)/2,

p2=Sqrt[−Discrim]}];

pp=p1*p1+p2*p2;

exp1:=Exp[p1*(tt−t0)];

cos1:=Cos[p2*(tt−t0)];

sin1:=Sin[p2*(tt−t0)];

ev1=(L1−C1*r1*Rn)/(2*L1*Rn);

dtrm=Sqrt[(L1+C1*r1*Rn)^2−4*C1*L1*Rn*(r1+Rn)]/(2*C1*L1*Rn);

In this cell, we determine the discriminant (denoted **Discrim**) of the characteristic equation corresponding to the transfer function of the filter and load. Then we define some parts of the solution to the differential equation. Using these parts, we define the currents and voltages as follows:

Iap:=1/(2*C1*dtrm)*(C1*dtrm*(Exp[p1*(tt−t0)]+Exp[p2*(tt−t0)])*I1+

C1^2*dtrm^2*(Exp[p1*(tt−t0)]−Exp[p2*(tt−t0)])*U1−

(Exp[p1*(tt−t0)]−Exp[p2*(tt−t0)])*ev1*(−I1+ev1*U1))−

1/(2*C1*dtrm*(r1+Rn))*(EL*((Exp[p1*(tt−t0)]−Exp[p2*(tt−t0)])*ev1*

L1+C1*(dtrm*(−2+Exp[p1*(tt−t0)]+Exp[p2*(tt−t0)])*

L1+(−Exp[p1*(tt−t0)]+Exp[p2*(tt−t0)])*Rn)));

Uap:=1/(2*C1*dtrm)*(Exp[p1*(tt−t0)]*(I1+(C1*dtrm−ev1)*U1)+

Exp[p2*(tt−t0)]*(−I1+(C1*dtrm+ev1)*U1))−

1/(2*C1*dtrm*(r1+Rn))*(EL*((Exp[p1*(tt−t0)]−Exp[p2*(tt−t0)])*ev1*

L1+C1*(dtrm*(−2+Exp[p1*(tt−t0)]+Exp[p2*(tt−t0)])*

L1+(Exp[p1*(tt−t0)]−Exp[p2*(tt−t0)])*r1))*Rn);

Ics:=(cos1+(p1−a22)/p2*sin1)*exp1*I1+a12/p2*sin1*exp1*U1+

a22/pp*cos1*exp1*EL+(pp−p1*a22)/p2/pp*sin1*exp1*EL−a22*EL/pp;

Ucs:=a21/p2*sin1*exp1*I1+(cos1+(p1−a11)/p2*sin1)*exp1*

U1+(−a21/pp*exp1*cos1+p1*a21/pp/p2*exp1*sin1)*EL+a21*EL/pp;

Ik:=If[Discrim>0,Iap,Ics];

Uk:=If[Discrim>0,Uap,Ucs];

Usl:=Uref*Sin[wf*tt+f0]–Uk;

The functions **Iap** and **Uap** define solutions to the current and voltage in the case of real roots (the descriminant **Discrim** is greater than zero); the functions **Ics** and **Ucs** define solutions to complex-conjugate roots. The functions **Ik** and **Uk** combine these solutions.

The procedure for the calculation of transient process is presented in the next cell:

Nm=5000;

t0=0;

U1=0;

I1=0;

E1=Abs[E1];

Km=40;

Tk=Tp/Km;

For[n=1,n<=Nm,n++,{For[m=1,m<=Km,m++,{tt=Tk*m+t0,

If[(((Usl<-pg) && (E1>0))||((Usl>pg) && (E1<0))),{t3=tt,t4=t3-Tk,

If[Usl<=-pg,Ftn=FindRoot[Usl==-pg,{tt,t4}],

Ftn=FindRoot[Usl==pg,{tt,t4}]],m=Km+1},1]}],

tt=Re[tt/.Ftn[[1]]],Ut=Uk,It=Ik,U1=Ut,Un[n]=Ut,I1=It,Iin[n]=It,t0=tt,

KT[n]=t0,E1=-E1}];

In this cell, **U1=0** and **I1=0** define the initial values of the voltage and current for the time **t0=0**, **Tk** defines the minimal part of the time interval when the switch instant **tt** is determined, **Ftn** defines the switching time, and **Un[n]** and **Iin[n]** define the vectors of voltages and currents calculated at the switching points.

The calculation procedure for the switching points of the relay element is based upon preliminarily finding a sufficiently small interval (in which the point is situated). For this, the total period is divided into **Km** intervals with a duration equal to **Tp/Km**. The number of points is given by the variable **Nm**. The calculation is realized for initial conditions equal to zero. From this, it is assumed that the relay element is in the state **+E1**. For the frequency 10000 Hz in the system, steady-state oscillations are formed. The plotting of the transient process for first 20 switches of the relay element is done using the **Zu1[n_]**, **Zu2[n_]**, and **Ris** functions:

$$B1a = \begin{pmatrix} Abs[E1]/L1 \\ 0 \end{pmatrix};$$

$$B2a = \begin{pmatrix} -Abs[E1]/L1 \\ 0 \end{pmatrix};$$

Clear[tt,n,Ris];

$$For[n=1,n<=Nm,n++, Xn[n] = \begin{pmatrix} Iin[n] \\ Un[n] \end{pmatrix};$$

KT[0]=0;

$$Xn[0] = \begin{pmatrix} 0 \\ 0 \end{pmatrix};$$

Zu1[n_]:=If[KT[n]<=tt<KT[n+1],Part[MatrixExp[A1*(tt-KT[n])].

Xn[n]+A1inv.(MatrixExp[A1*(tt-KT[n])]-I2).B1a,2],0];

Zu2[n_]:=If[KT[n+1]<=tt<KT[n+2],Part[MatrixExp[A1*(tt-KT[n+1])].

Xn[n+1]+A1inv.(MatrixExp[A1*(tt-KT[n+1])]-I2).B2a,2],0];

Nint=8;

Ris=Sum[Zu1[n]+Zu2[n],{n,0,Nint,2}];

Plot[{Uref*Sin[wf*tt+f0],Ris},{tt,KT[0],KT[Nint+2]},AxesLabel->{"t","u"}];

In this cell, **Nint** defines the number of intervals used for the summing of solutions. The functions **Zu1[n_]** and **Zu2[n_]** define the solution to even and odd time intervals, respectively, situated between the two switching points of the relay element.

The graphs of the transient process for the voltage across the capacitor for the first 10 switches of the relay element are presented in Figure 3.37. The graph of the transient process for the last 10 switches (Figure 3.38) is made with the help of the functions

ZU[tt_,n_]:=If[KT[n]<=tt<KT[n+1],Part[MatrixExp[A1*(tt-KT[n])].

Xn[n]+A1inv.(MatrixExp[A1*(tt-KT[n])]-I2).B1,2],0]+

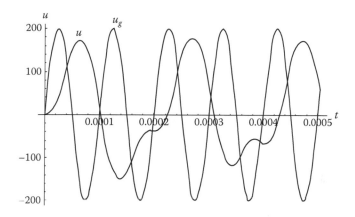

FIGURE 3.37
The transient process for voltage u across the capacitor and the generator voltage u_g for the first 10 switches Ff = 10000 Hz (u in volts, time t in seconds).

If[KT[n+1]<=tt<KT[n+2],Part[MatrixExp[A1*(tt-KT[n+1])].

Xn[n+1]+A1inv.(MatrixExp[A1*(tt-KT[n+1])]-I2).B2,2],0];

FU2[tt_]:=Sum[ZU[tt,n],{n,Nm-10,Nm-2,2}];

Plot[{Uref*Sin[wf*tt+f0],FU2[tt]},{tt,KT[Nm-10],KT[Nm]},

AxesLabel->{"t","u"},AxesOrigin->{KT[Nm-10],0}];

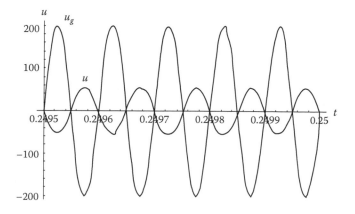

FIGURE 3.38
The transient process for voltage u across the capacitor and the generator voltage u_g for the last 10 switches Ff = 10000 Hz (u in volts, time t in seconds).

In this cell, the function **ZU[tt_,n_]** combines functions **Zu1[n_]** and **Zu2[n_]**. The stability calculation

$$Ta=1.0/Ff/2;$$

$$tn=ArcSin[(Part[Part[Xt1,2],1]-pg)/Uref];$$

$$Udt1=Part[Part[A1.Xt1,2],1];$$

$$f12s=-2*Abs[EL]*Ky*Kd/Abs[Ky*Kd*Uref*wf*Cos[(Pi-tn)]-Kd*Udt1];$$

$$Abs[Sei]$$

gives the absolute values of eigenvalues

$$\{0.976937, 0.976937\}.$$

Let us consider the behavior of the relay system for the generator frequency equal to 4636 Hz. Graphs of the transient process for the first and last 10 switches of the relay element are presented in Figures 3.39 and 3.40.

Figure 3.40 shows that, in the system, on the interval equal to approximately 2500T, a steady-state process with the period of external supply is not formed. Confirmation of this fact is shown by the stability calculation. One of the absolute values of the eigenvalues is greater than unity:

$$\{1.55916, 0.579952\}$$

Consider the characteristics of process identification. For this, we plot a phase-space portrait. In order to simplify the plotting procedure, we use

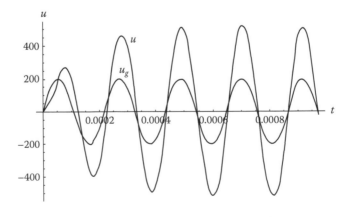

FIGURE 3.39
The transient process for voltage u across the capacitor and the generator voltage u_g for the first 10 switches FF = 4636 Hz (u in volts, time t in seconds).

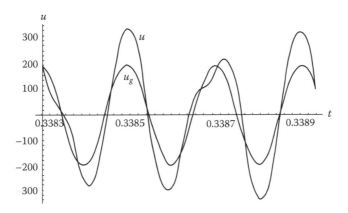

FIGURE 3.40
The transient process for voltage u across the capacitor and the generator voltage u_g for the last 10 switches FF = 4636 IIz (u in volts, time t in seconds).

the values **Un[2*n]**, **Iin[2*n]**, and **Un[2*n-1]**, **Iin[2*n-1]** determined in the switching moments of the relay element:

uiTab=Table[{Un[2*n],Iin[2*n]},{n,1000,Round[Nm/2]}];

LPeven=ListPlot[uiTab,AxesLabel->{"u","i"},DisplayFunction->Identity];

uiTab=Table[{Un[2*n-1],Iin[2*n-1]},{n,1000,Round[Nm/2]}];

LPodd=ListPlot[uiTab,AxesLabel->{"u","i"},DisplayFunction->Identity];

Show[{LPeven,LPodd},DisplayFunction->$DisplayFunction];

The phase-plane portrait is shown in Figure 3.41. The Poincare section $u_{2(n+1)} = f(u_{2n})$ is presented in Figure 3.42. The plotting of the Poincare section is realized for the interval **200 ... Nm**

uTab=Table[{Un[2*(n-1)],Un[2*n]},{n,100,Round[Nm/2]}];

ListPlot[uTab,AxesLabel->{"Un","Un+1"}]

The images presented in Figures 3.41 and 3.42 indicate that quasi-periodic oscillations are formed in the system.
     We also make the correlation function for the even-switching moments:

Nt=400;

Nmm=Round[Nm/2];

Xav=Sum[Un[2*k],{k,1,Nmm}]/Nmm;

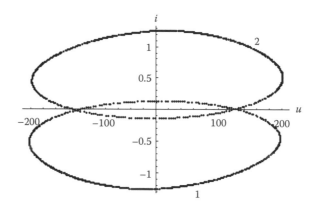

FIGURE 3.41
The phase-plane portrait for the switching moments (*i* in amperes, *u* in volts): the curve (1) is obtained for even-switching moments, and curve (2) for odd-switching moments.

For[n=1,n<=Nt,n++,Kor[n]=Sum[(Un[2*i]-Xav)*(Un[2*i+2*n]-Xav),

{i,1,Nmm-Nt}]/(Nmm-Nt)];

TabKor=Table[{i,Kor[i]},{i,Nt}];

ListPlot[TabKor,Prolog->AbsolutePointSize[2],AxesLabel->{"m","K(m)"}];

The variable **Nt** defines the number of points in which the correlation function is calculated, the variable **Nmm** defines the number of even-switching moments, the variable **Xav** defines the average value of even-switching moments, and the **Prolog** option allows increase in the size of points in a graph. The correlation function is shown in Figure 3.43. The figure illustrates the regularity of the correlation function. This confirms that the process is

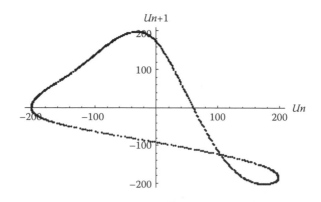

FIGURE 3.42
The Puancare section of the voltage for even-switching moments $u_{2n+2} = f(u_{2n})$ (U_{n+1} and U_n in volts).

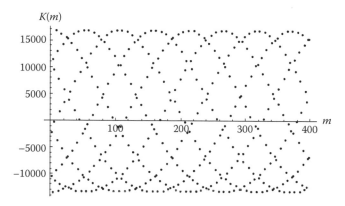

FIGURE 3.43
The correlation function for even-switching moments of the relay element.

quasi-periodic. It should be noted that the correlation function for all switching moments does not have a regular structure.

Consider a particularity of process identification done with the help of the space dimension calculation:

$$\text{Clear[k,n,m];}$$

$$\text{rk=0.2;}$$

$$\text{Nk=100;}$$

$$\text{Nmin=Nmm-100;}$$

$$\text{Norma:=Sqrt[(Un[2*n]-Un[2*m])^2+(Iin[2*n]-Iin[2*m])^2];}$$

$$\text{For[k=1,k<=Nk,k++,\{Cr[k]=0;rr=rk*k;For[m=Nmin,m<=Nmm,m++,}$$

$$\text{For[n=Nmin,n<=Nmm,n++,If[(rr>=Norma),Cr[k]=Cr[k]+1,1]]\}];}$$

$$\text{TabP=Table[\{Log[rk*i],Log[Cr[i]/((Nmm-Nmin)^2)]\},\{i,Nk\}];}$$

$$\text{ListPlot[TabP,Prolog->AbsolutePointSize[4],AxesLabel->\{"lnr","lnC(r)"\}]}$$

The variable **rk** defines the minimum dimension of the cell in the phase space; the variable **Nk** defines the number of calculating points, the variable **Norma** denotes the distance between points, and the function **For[ ]** is used for the calculation of the correlation function as in (3.55). The graph of the function ln $C(r)$ versus ln r is presented in Figure 3.44. This graph is not uniform. The confirmation of nonuniformity is illustrated in the graph obtained by the calculation of the derivative

$$\frac{\partial \ln C(r)}{\partial \ln r}$$

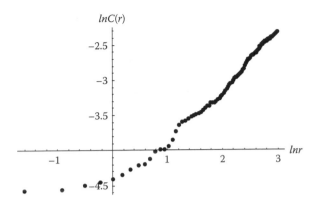

FIGURE 3.44
The graph of the function ln $C(r)$ versus ln r.

In the cell, this derivative is calculated:

TabCr=Table[{Log[rk*i],(Log[Cr[i+1]]−

Log[Cr[i]])/((Nmm-Nmin)^2)/(Log[rk*(i+1)]-Log[rk*i])},{i,Nk-1}];

ListPlot[TabCr,Prolog->AbsolutePointSize[4],AxesLabel->{"lnr","lnC(r)/lnr"}]

The graph of the function $\frac{\partial \ln C(r)}{\partial \ln r}$ versus ln r is presented in Figure 3.45. It should be noted that the dependence presented in Figure 3.45 has certain maximums.

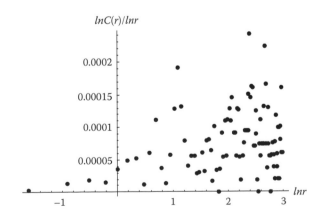

FIGURE 3.45
The graph of the function $\frac{\partial \ln C(r)}{\partial \ln r}$ versus ln r.

The harmonic linearization method discussed here expediently uses for analysis a closed-loop system in which an input forcing is absent. In this case, the results obtained by the method practically coincide with simulation ones. If, on the input of a system, a sinusoidal voltage is presented, the stability of the periodic and subharmonic oscillations present are expediently analyzed with the help of the linearization method.

4

Analysis of Processes in Systems with Converters

4.1 Power Conditioner

4.1.1 The Mathematical Model of a System

An AC converter is used as a power conditioner, a compensator for the sag or imbalance of voltage in power supply, and a compensator for reactive power. In such converters, control methods are used, providing the possibility of dynamic change of the transformation ratio with a time constant that is much less than the period of the supply voltage (Veszpremi and Hunyar, 2000; Kasperek, 2003).

Consider a mathematical model of the power conditioner, the circuits of which are constructed on the basis of the Buck topology (Figure 4.1).

This power conditioner provides direct conversion of AC voltage without an intermediate circuit used for energy storage. In the system, a voltage imbalance is introduced by the connection of the resistor R_n.

Assume that the switches are described by the RS model, and the inductors and load are linear. Then the electromagnetic processes for the interval when switches S_{1s} and S_{2s} are closed, and switches S_{1L} and S_{2L} are opened, are described by the matrix differential equation (Korotyeyev and Kasperek, 2004a)

$$LL\frac{dI}{dt} = -A_{11}I - AI_{11}i - AU_{11}I_0 + E, \tag{4.1}$$

where

$$LL = \begin{vmatrix} L_{L1} + L_{L2} & L_{L2} \\ L_{L2} & L_{L2} + L_{L3} \end{vmatrix};$$

$$I = \begin{vmatrix} i_{L1} \\ i_{L3} \end{vmatrix}; \quad I_0 = \begin{vmatrix} i_{01} \\ i_{03} \end{vmatrix};$$

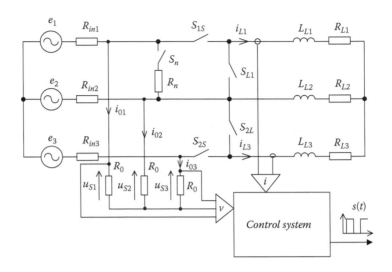

FIGURE 4.1
System topology with power conditioner. (Data from Korotyeyev I. Ye. and Kasperek R., 2004a. With permission.)

$$A_{11} = \begin{vmatrix} R_{L1} + R_{L2} + & \\ & R_{L2} + R_{in2} \\ R_{in1} + R_{in2} & \\ & R_{L2} + R_{L3} + \\ R_{L2} + R_{in2} & \\ & R_{in2} + R_{in3} \end{vmatrix} ;$$

$$AI_{11} = \begin{vmatrix} R_{in1} + R_{in2} \\ R_{in2} \end{vmatrix} ;$$

$$AU_{11} = \begin{bmatrix} R_{in1} + R_{in2} & R_{in2} \\ R_{in2} & R_{in2} + R_{in3} \end{bmatrix} ;$$

$$E = \begin{bmatrix} e_1 - e_2 \\ e_2 - e_3 \end{bmatrix} .$$

Suppose that the switch S_n is in the on state. Then the current is

$$i = i_{11} + RD_{11}I_0 + RP_{11}I ; \tag{4.2}$$

where

$$i_{11} = \frac{e_1 + e_2}{R_s} ; \quad R_s = R_{in1} + R_{in2} + R_n ; \quad RD_{11} = RP_{11} = \begin{bmatrix} -\dfrac{R_{in1} + R_{in2}}{R_s} & \dfrac{R_{in2}}{R_s} \end{bmatrix} .$$

The algebraic equation for the balancing circuit has the form

$$E = AD_{11}I_0 + AN_{11}i + AP_{11}I; \tag{4.3}$$

where

$$AD_{11} = \begin{vmatrix} R_{in1} + R_{in2} + 2R_0 & R_{in2} + 2R_0 \\ R_{in2} + 2R_0 & R_{in2} + R_{in3} + 2R_0 \end{vmatrix}; \quad AN_{11} = AI_{11};$$

$$AP_{11} = \begin{vmatrix} R_{in1} + R_{in2} & R_{in2} \\ R_{in2} & R_{in2} + R_{in3} \end{vmatrix}.$$

The electromagnetic processes for the interval when switches S_{1L} and S_{2L} are closed, and switches S_{1s} and S_{2s} are opened, are described by the matrix differential equation

$$LL\frac{dI}{dt} = -A_{22}I; \tag{4.4}$$

$$i = i_{11} + RD_{11}I_0; \tag{4.5}$$

$$E = AD_{11}I_0 + AN_{11}i, \tag{4.6}$$

where

$$A_{22} = \begin{vmatrix} R_{L1} + R_{L2} & R_{L2} \\ R_{L2} & R_{L2} + R_{L3} \end{vmatrix}.$$

Combining Equations 4.1–4.3 and 4.4–4.6, we obtain

$$LL\frac{dI}{dt} = -A_{22}I - \gamma AP_{11}I - \gamma AU_{11}I_0 - \gamma AI_{11}i + \gamma E \tag{4.7}$$

$$i = i_{11} + RD_{11}I_0 + \gamma RP_{11}I; \tag{4.8}$$

$$E = AD_{11}I_0 + AN_{11}i + \gamma AP_{11}I, \tag{4.9}$$

where γ is the switching function for the switches S_s and S_L.

The control system presented in Figure 4.2 generates impulses based on the calculation of instantaneous power. During the calculation process, the Clark transformation is used.

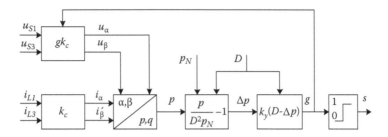

FIGURE 4.2
Schematic diagram of the control system. (Data from Korotyeyev I. Ye. and Kasperek R., 2004a.
With permission.)

The processes in the control system are described by the following equations:

$$I_\alpha = k_c I;$$

$$U_\alpha = k_c U;$$

$$P_\alpha = g U_\alpha^T I_\alpha;$$

$$\Delta p = \frac{P_\alpha}{D^2 p_n} - 1; \qquad (4.10)$$

$$g = k_y (D - \Delta p);$$

$$u_{co} = g u_g(t);$$

$$s = s(u_{co}),$$

where

$$I_\alpha = \begin{vmatrix} i_\alpha \\ i_\beta \end{vmatrix}; \quad U_\alpha = \begin{vmatrix} u_\alpha \\ u_\beta \end{vmatrix}; \quad U = \begin{vmatrix} u_{S1} \\ u_{S3} \end{vmatrix}; \quad k_c = \begin{bmatrix} 1 & 0 \\ 1 & 2 \\ \sqrt{3} & \sqrt{3} \end{bmatrix};$$

D, p_n are constancies; k_y is the gain; $u_g(t)$ is the voltage of the independent generator; and U_α^T is the transposed vector.

Since $|AP_{11}| < |A_{22}|$, $|AU_{11}| < |A_{22}|$, $|AP_{11}I| < |AN_{11}i|$ and $|RP_{11}I| < i_{11}$, Equation 4.7 is simplified as

$$LL \frac{dI}{dt} = -A_{22} I + \gamma E_1, \qquad (4.11)$$

where $E_1 = E - AI_{11}i_{11} - (AI_{11}RD_{11} + AU_{11})I_0$; $I_0 = RS^{-1}(E - AN_{11}i_{11})$; $RS = AD_{11} + AN_{11}RD_{11}$; and RS_{11}^{-1} is the inverse matrix. The voltage U is defined by the expression

$$U = R_0 I_0. \tag{4.12}$$

Using the averaged state-space method (Middlebrook and Ćuk, 1976; Korotyeyev and Fedyczak, 2002), we transform Equations 4.10 and 4.11 into the averaged state-space equations

$$LL\frac{d\overline{I}}{dt} = -A_{22}\overline{I} + dE_1; \tag{4.13}$$

$$U = R_0 I_0;$$

$$\overline{I}_\alpha = k_c \overline{I};$$

$$U_\alpha = k_c U;$$

$$\left. \begin{array}{c} d = \dfrac{g}{U_{ag}}; \\[2ex] g = \dfrac{k_y(D+1)}{1 + k_y k_d U_\alpha^T I_\alpha} \end{array} \right\}, \tag{4.14}$$

where $\overline{I}$, and $\overline{I}_\alpha$ are the averaged vectors; d is the duty factor; U_{ag} is the amplitude of the generator voltage; and $k_d = \frac{1}{D^2 p_n}$.

4.1.2 Computation of a Steady-State Process

Since in the control system signals are generated on the basis of the calculation of power, we will find the solving of an equation set to be the sum of the constant component and second harmonic. We assume that the duty factor is

$$d = d_0 + d_2 \sin(2\omega t + \varphi), \tag{4.15}$$

where d_0 is the constant component and d_2 and φ are the amplitude and phase, respectively, of the second harmonic. Using the expression (4.15), we determine the solution to the set (4.13) and (4.14) by taking the Laplace transform

$$I(p) = (pI - LL^{-1}A_{22})^{-1}LL^{-1}d_0E(p) + d_2(pI - LL^{-1}A_{22})^{-1}LL^{-1} \cdot$$

$$\left[E(p) * \frac{p\sin\varphi + 2\omega\cos\varphi}{p^2 + (2\omega)^2} \right], \tag{4.16}$$

where $I(p)$ and $E(p)$ are the transforms of the vectors $\overline{I}$ and E_1, respectively; I is the unity matrix; $(..)^{-1}$ and LL^{-1} are inverse matrixes; and $*$ is the convolution in the complex domain.

We can write the steady-state current as follows:

$$\overline{I} = I_0(t) + I_2(t), \tag{4.17}$$

where

$$I_0(t) = 2\operatorname{Re}\left[(j\omega I - LL^{-1}A_{22})^{-1} \frac{d_0}{2j\omega} LL^{-1}E_n(j\omega)e^{j\omega t} \right] \tag{4.18}$$

is determined with respect to the poles $p = \pm j\omega$ of the vector $E(p)$. In this expression, $E_n(j\omega)$ denotes the numerator of the vector $E(j\omega)$.

Consider the second term in the solution (4.16). Calculating the convolution of the function in (4.16), we obtain

$$E_c(p) = E(p) * \frac{p\sin\varphi + 2\omega\cos\varphi}{p^2 + (2\omega)^2} = \sum_{\substack{k=-1 \\ k\neq 0}}^{1} \frac{(p - kj\omega)\sin\varphi + 2\omega\cos\varphi}{(p + kj\omega)(p - kj3\omega)} \frac{E_n(kj\omega)}{k2j\omega}.$$

From the expression, it follows that, in the solution, the first and third harmonics are presented. Since a third harmonic does not participate in the solution, from this expression we extract only a first harmonic

$$E_{c1}(p) = \sum_{\substack{k=-1 \\ k\neq 0}}^{1} \frac{-2kj\omega\sin\varphi + 2\omega\cos\varphi}{(p + kj\omega)(-kj4\omega)} \frac{E_n(kj\omega)}{k2j\omega}.$$

Then the value of the vector of the current $I_2(t)$ in (4.17) is determined by the expression similar to (4.18):

$$I_2(t) \cong 2\operatorname{Re}\left[(j\omega I - LL^{-1}A_{22})^{-1} \frac{d_2}{2j\omega} LL^{-1}E_{c1n}(j\omega)e^{j\omega t} \right], \tag{4.19}$$

where $E_{c1n}(j\omega)$ is the numerator of the vector $E_{c1}(j\omega)$.

The instantaneous power

$$P_\alpha = U_\alpha^T \overline{I}_\alpha \tag{4.20}$$

can be written as

$$P_\alpha = P_0 + P_2 \sin(2\omega t + \psi), \qquad (4.21)$$

where $P_0 = \frac{1}{T}\int_0^T P_\alpha dt$ is the constant component of the instantaneous power; and P_2 and ψ are the amplitude and phase, respectively, of the second harmonic.

Using the Laplace transform, we calculate the power (4.20) with the help of the convolution

$$P_\alpha(p) = U_\alpha^T(p) * I_\alpha(p) = \sum_{\substack{k=-1 \\ k\neq 0}}^{1} \frac{U_{\alpha n}^T(kj\omega)I_{\alpha n}(p - jk\omega)}{2kj\omega[(p - jk\omega)^2 + \omega^2]} = \sum_{\substack{k=-1 \\ k\neq 0}}^{1} \frac{U_{\alpha n}^T(kj\omega)I_{\alpha n}(p - jk\omega)}{2kj\omega p(p - 2jk\omega)},$$

$$(4.22)$$

where $I_{\alpha n}(kj\omega)$ and $U_{\alpha n}^T(kj\omega)$ are the numerators of the vectors $I_\alpha(kj\omega)$, $U_\alpha^T(kj\omega)$, respectively.

From the expression (4.22), it follows that, in the solution, a constant component and second harmonic exist. We determine the constant component using the *theorem of the final value* of the Laplace transform

$$P_0 = \lim_{p \to 0} p P_\alpha(p) = \sum_{\substack{k=-1 \\ k\neq 0}}^{1} \frac{U_{\alpha n}^T(kj\omega)I_{\alpha n}(-jk\omega)}{4k^2\omega^2}.$$

The transformation of the second harmonic is determined by the calculation of residues

$$P_2(p) = \sum_{\substack{k=-1 \\ k\neq 0}}^{1} \frac{\lim\limits_{p \to 2jk\omega}(p - j2k\omega)P_\alpha(p)}{p - j2k\omega} = P_2 \frac{p \sin \psi + 2\omega \cos \psi}{p^2 + 4\omega^2}.$$

In this expression, P_2 and ψ are determined as a result of the calculation of a limit of the transform $P_\alpha(p)$.

Taking into account that the power P_2 is small in comparison to P_0, we expand (4.14) in the Taylor series about P_0. Then, the duty factor d takes form

$$d = \frac{k_y(D+1)}{(1 + k_y k_d P_0)U_{ag}} - \frac{k_y(D+1)k_y k_d P_2 \sin(2\omega t + \psi)}{(1 + k_y k_d P_0)^2 U_{ag}}.$$

Using the expression for the duty factor (4.15), we extract the equation for the constant component:

$$d_0 = \frac{k_y(D+1)}{(1+k_yk_dP_0)U_{ag}}$$ (4.23)

and the equation for the amplitude d_2 and phase φ:

$$d_2 = \frac{k_y(D+1)k_yk_dP_2}{(1+k_yk_dP_0)^2U_{ag}} \; ;$$ (4.24)

$$\psi = \varphi + \pi.$$ (4.25)

Transform Equation 4.23 in the following way:

$$d_0 = \frac{k_y(D+1)}{(1+k_yk_dd_0\tilde{P}_0)U_{ag}} ,$$

where $\tilde{P}_0 = \frac{P_0}{d_0}$; and $\tilde{P}_0 = \frac{1}{T}\int_0^T U_\alpha^T I_{0\alpha}dt$; $I_{0\alpha} = k_cI_0(t)$. Then the constant component is determined by the solution to the square equation

$$d_0U_{ag}(1+k_yk_dd_0\tilde{P}_0) = k_y(D+1).$$ (4.26)

For the determination of d_0, and d_2 φ, at first we calculate d_0 from Equation 4.26, and then we solve Equations 4.24 and 4.25 simultaneously.

4.1.3 Steady-State Stability Analyses

For the purpose of calculation of the stability of steady-state behavior, we find increments of the state variables. Using Equations 4.13 and 4.14, we obtain (Korotyeyev and Kasperek, 2004b)

$$LL\frac{d\bar{I}_\xi}{dt} = -A_{22}\bar{I}_\xi + d_\xi E_1 ;$$ (4.27)

$$\bar{I}_{\alpha\xi} = k_c\bar{I}_\xi ;$$

$$d_\xi = -\frac{k^2k_r(D+1)U_\alpha^T}{U_{ag}(1+kk_rP_{\alpha s})^2}\bar{I}_{\alpha\xi} ,$$ (4.28)

where $\bar{I}_\xi$ and d_ξ are the variations of the variables $\bar{I}$ and d, respectively; and $P_{\alpha s}$ is the value of P_α for the steady-state behavior. Substituting (4.28) into (4.27), we obtain the linearized equation

$$LL\frac{d\bar{I}_\xi}{dt} = -M\bar{I}_\xi ,$$ (4.29)

where

$$M = A_{22} + E_1 \frac{k^2 k_r (D+1) U_\alpha^T k_c}{U_{ag}(1 + kk_r P_{as})^2}.$$

Equation 4.29 is a differential equation with periodic coefficients. For stability analysis, it is necessary to determine the solution to this equation for the interval equal to the period of the matrix M. We find the solution by approximating E_1, U_α^T, and P_{as} at a small time interval with constant values. Then, supposing that E_1, U_α^T, and P_{as} are constants for the interval $t_i - t_{i+1}$, we can write the solution to Equation 4.29 as follows:

$$\overline{I}_\xi = e^{-M(t_i)(t-t_i)} \overline{I}_\xi(t_i),$$

where $e^{-M(t_i)(t-t_i)}$ is the matrix exponential. Solving the equation for all intervals and eliminating intermediate variables, we determine the solution for the period:

$$\overline{I}_\xi((n+1)T) = \prod_{i=1}^{N} e^{-M(t_i)\tau} \overline{I}_\xi(nT), \tag{4.30}$$

where $\tau = t_{i+1} - t_i = const$; $N = \frac{T}{\tau}$. Analysis of system stability is based on the calculation of the eigenvalues of the matrix

$$H = \prod_{i=1}^{N} e^{-M(t_i)\tau}. \tag{4.31}$$

The system is stable if all absolute values of the eigenvalues of the matrix H are less than unity.

4.1.4 Calculation of Steady-State Processes and System Stability

Let us calculate a steady-state process for the following values of the parameters: $E = 310$ V; $D = 0.5$; $k_u = 0.0326$; $k_i = 3.256$; $P_n = 97.5$ W; $U_{ag} = 1$ V; $R_{in} = 1\ \Omega$, $R_n = 10\ \Omega$, $R_L = 100\ \Omega$; $L_L = 75$ mH; $T = 20 \cdot 10^{-3}$ s (the switching period of the converter). Coefficients k_u and k_i are used for the calculation of the vectors

$$\overline{I}_\alpha = k_i k_c \overline{I},$$

$$U_\alpha = k_u k_c U.$$

Let us use Mathematica® for a process analysis in the system. In the cell, the value of the parameters are assigned and matrixes are defined

$$L1=L2=L3=0.075;$$

$$r1=r2=r3=100;$$

$$re1=re2=re3=1;$$

$$R12=10;$$

$$RR=re1+re2+R12;$$

$$ri=10*(10^3);$$

$$Es=310;$$

$$pzn=97.5;$$

$$ky=10;$$

$$D1=0.5;$$

$$ku=0.0326;$$

$$ki=3.256;$$

$$Ff=50.0;$$

$$T=5.0*10^3;$$

$$Kt3F=Round[1/Ff/T];$$

$$kc = \begin{pmatrix} 1 & 0 \\ 1/\sqrt{3} & 2/\sqrt{3} \end{pmatrix};$$

$$LL = \begin{pmatrix} L1+L2 & L2 \\ L2 & L2+L3 \end{pmatrix};$$

$$Clear[e1,e2,e3,d1];$$

$$EE := \begin{pmatrix} e1-e2 \\ e3-e2 \end{pmatrix};$$

$$A11 = \begin{pmatrix} r1+re1+r2+re2 & r2+re2 \\ r2+re2 & r2+re2+r3+re3 \end{pmatrix};$$

$$AV11 = \begin{pmatrix} re1 + re2 & re2 \\ re2 & re2 + re3 \end{pmatrix};$$

$$AI11 := \begin{pmatrix} re1 + re2 \\ re2 \end{pmatrix};$$

$$AD11 = \begin{pmatrix} re1 + re2 + 2 * ri & re2 + ri \\ re2 + ri & re2 + re3 + 2 * ri \end{pmatrix};$$

$$AP11 := \begin{pmatrix} re1 + re2 & re2 \\ re2 & re2 + re3 \end{pmatrix};$$

$$AN11 := \begin{pmatrix} re1 + rc2 \\ re2 \end{pmatrix}$$

RD11:=(-(re1+re2)/RR -re2/RR);

I11=(e1-e2)/RR;

RP11:=(-(re1+re2)/RR -re2/RR);

I2=IdentityMatrix[2];

The coefficient **Kt3F** defines the number of switches during the period of power supply.
   In the next cell, intermediate matrixes and vectors are calculated:

RS11=AD11+AN11.RD11;

RS11inv=Inverse[RS11];

Uri=Simplify[N[(RS11inv.(EE-AN11*I11))*ri]]

E11=EE11-AI11*I11-(AI11.RD11+AV11).RS11inv.(EE-AN11*I11);

Ev=Simplify[N[E11*d1]]

Linv=Inverse[LL];

In this cell, U_{ri} denotes U, and E_v denotes dE_1.
   We define the phase voltages and calculate the Laplace transform of the voltages E_1 and U_p as follows:

Clear[ω];

e1:=Es*Sin[ω *t];

e2:=Es*Sin[ω*t-2*Pi/3];

e3:=Es*Sin[ω *t+2*Pi/3];

Up=Simplify[LaplaceTransform[Uri,t,p]];

Ep=Simplify[LaplaceTransform[Ev,t,p]]

Expressions that correspond to numerators of transforms U_p and E_p are determined in the cell:

Clear[d1];

Uω=Numerator[Up]/.p->I*ω;

Umω=Numerator[Up]/.p->I*ω;

Epω=Numerator[Ep]/.p->(p-I*ω);

Epmω=Numerator[Ep]/.p->(p+I*ω);

In the cell, the convolution $E(p) * \frac{p\sin\varphi + 2\omega\cos\varphi}{p^2 + (2\omega)^2}$ entered in expression (4.16) is calculated, and the first and third harmonics denoted by **d2Es1** and **d2Es3** are extracted from the obtained expression:

Clear[ω,d1];

Snom=LaplaceTransform[Sin[2*ω*tt+φ],tt,p];

d2Es=Chop[FullSimplify[Limit[(q-I*ω)*(Snom/.p->(p-q))*(Ep/.p->q)/ d1,q->I*ω]+

Limit[(q+I*ω)*(Snom/.p->(p-q))*(Ep/.p->q)/d1,q->-I*ω]]];

d2Es1=Simplify[Limit[(p-I*ω)*d2Es,p->I*ω]/(p-I*ω)+

Limit[(p+I*ω)*d2Es,p->-I*ω]/(p+I*ω)];

d2Es3=Chop[Simplify[Limit[(p-3*I*ω)*d2Es,p->3*I*ω]/(p-3*I*ω)+

Limit[(p+3*I*ω)*d2Es,p->-3*I*ω]/(p+3*I*ω)]];

For harmonics determination we use the function. It should be recalled that the third harmonic does not take part in the solution. In what follows, this harmonic is used for plotting currents.

Now we will calculate components of the power. First we calculate the part of the power that corresponds to the current $I_2(t)$. This part of the power is determined by calculating a convolution for the first harmonic of the voltage and current. We use the expression $E_{c1}(p)$ and a transform for the current $I_2(t)$.

m1s=Simplify[Transpose[kc.(Up/.p->q)].kc.Inverse[(p−q)*I2+Linv.A11].

Linv.(d2Es1/.p->(p−q))];

Sd12S=Simplify[Limit[(q−I*ω)*m1s,q->I*ω]+Limit[(q+I*ω)*m1s,q->I*ω]];

From the obtained expression we determine the part of the power P_2 caused by the current $I_2(t)$, and the sine and cosine components for that power:

Clear[ω];

Gd2=Simplify[Part[Part[Limit[(p−I*2*ω)*Sd12S,p->I*2*ω],1],1]/(p−I*2*ω)
+Part[Part[Limit[(p+I*2*ω)*Sd12S,p->I*2*ω],1],1]/(p+I*2*ω)];

ω =2*Pi*Ff;

Gd2N=Collect[ComplexExpand[Re[Numerator[Simplify[ki*ku*Gd2]]]],p]

cs2=Extract[Gd2N,{1}]/(2*ω)+Extract[Gd2N,{2}]/(2*ω);

sn2=Extract[Gd2N,3]/p;

In this cell, **cs2** and **sn2** denote the cosine and sine components of the power P_2.

Further, we determine the part of the power caused by the current $I_0(t)$. From this part the power P_0 is determined, and then the component value d_0 is calculated:

Sd1=Simplify[Transpose[kc.Uω].kc.Inverse[(p−I*ω)*I2+

Linv.A11].Linv.Epω (2*I*ω)/p/(p−I*2*ω)];

Sd1m=Simplify[Transpose[kc.Umω].kc.Inverse[(p+I*ω)*I2+

Linv.A11].Linv.Epmω/(−2*I*ω)/p/(p+I*2*ω)];

Clear[ω];

ω=2*Pi*Ff;

Ss1=Sd1/d1;

pcon=Part[Part[2*Re[N[Ss1*p/.p->0]]*ki*ku,1],1]

Ss1m=Sd1m/d1;

kd=1/(D1*D1*pzn);

Xs=Solve[(ky*kd*pcon)*x^2+x−ky(D1+1)==0,x]

x2=x/.x->Part[Xs,2];

d0=x/.Part[x2,1]

As a result of the calculation of the square equation (4.26), denoted **Xs** one obtains

$$\{\{x\text{->}-0.686487\}, \{x\text{->}0.656444\}\}$$

Since the constant component d_0 can only be positive, its value equals

$$0.656444$$

In the cell, another part of the power P_2 caused by the current $I_0(t)$ and the sine and cosine components for that power are determined:

Sd1S=Simplify[Ss1+Ss1m];

Gd1=Simplify[Part[Part[Limit[(p-I*2*ω)*Sd1S,p->I*2*ω],1],1]/(p-I*2*ω)+

Part[Part[Limit[(p+I*2*ω)*Sd1S,p->I*2*ω],1],1]/(p+I*2*ω)];

Gd1N=ComplexExpand[Re[Numerator[Simplify[ki*ku*Gd1]]]]

cs1=Extract[Gd1N,1]/(2*ω)

sn1=Extract[Gd1N,2]/p

Using the obtained expressions, the square of amplitude of the power P_2 is calculated:

snd12=sn1*d1+sn2*d2;

csd12=cs1*d1+cs2*d2;

dd2=Factor[snd12^2+csd12^2]/.d1->d0

Solving Equations 4.24 and 4.25, we find the amplitude d_2 and phase φ:

eq1=(D1+1)*ky*(ky*kd)/((1+d0*ky*kd*pcon)^2)*Sqrt[dd2/.d1->d0];

des1=FindRoot[{eq1==d2,(ArcTan[csd12/.d1->d0,snd12/.d1->d0])==Pi+φ},

{d2,0.1},{φ,-0.5}];

eqd2=d2/.Part[des1,1];

Print["d2 = ",eqd2];

eqφ=φ/.Part[des1,2];

Print["φ = ",eqφ;]

The result of the calculations is to output the amplitude and phase values

$$d2 = 0.0784066$$

$$\varphi = -0.677469$$

For an accuracy estimation of the considered method, we will use a numerical calculation of the differential equations:

Clear[t,d1];

Uri:=Simplify[N[(RS11inv.(EE-AN11*I11))*ri]];

$$pxy:=Simplify[Part[Part[ku*ki*Transpose[kc.\begin{pmatrix} y[t] \\ x[t] \end{pmatrix}].kc.Uri,1],1]];$$

d1:=ky*(D1+1)/(1+ky*kd*pxy);

$$U1=Part[Part[LL.\begin{pmatrix} y'[t] \\ x'[t] \end{pmatrix},1],1]+Part[Part[A11.\begin{pmatrix} y[t] \\ x[t] \end{pmatrix},1],1]-$$
$$Part[Part[Ev,1],1];$$

$$U2=Part[Part[LL.\begin{pmatrix} y'[t] \\ x'[t] \end{pmatrix},2],1]+Part[Part[A11.\begin{pmatrix} y[t] \\ x[t] \end{pmatrix},2],1]-$$
$$Part[Part[Ev,2],1];$$

sol=NDSolve[{U1==0,U2==0,x[0]==y[0]==0},{x,y},{t,8/Ff}]

The process calculation is made for the interval {0,8/Ff}.

Plotting the graphs of the duty factor for the considered and numerical methods are done as follows:

prdPlot=Plot[d0+eqd2*Sin[2*ω*tt+eqφ],{tt,7/Ff,8/Ff},DisplayFunction-> Identity];

ndPlot=Plot[Evaluate[d1n/.sol],{t,7/Ff,8/Ff},DisplayFunction->Identity];

Show[{prdPlot,ndPlot},DisplayFunction->$DisplayFunction];

Temporal changes in the duty factor calculated by the numerical and considered methods are shown in Figure 4.3.

Let us calculate the current in the load. The current is found with the help of the inverse Laplace transform:

Clear[p,t];

Imax=Simplify[InverseLaplaceTransform[d0*

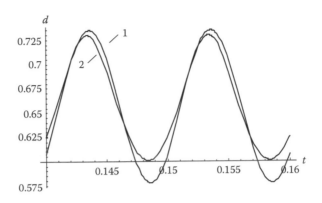

FIGURE 4.3
Temporal changes in the duty factor for the numerical (1) and considered (2) methods (time t in seconds).

Inverse[p*I2+Linv.A11].Linv.Ep/d1,p,t]];

Imax2=Simplify[InverseLaplaceTransform[eqd2*Inverse[p*I2+Linv.A11].

Linv.d2Es1,p,t]/.φ->eqφ];

Imax3=Simplify[InverseLaplaceTransform[eqd2*

Inverse[p*I2+Linv.A11].Linv.d2Es3,p,t]/.φ->eqφ];

The **Imax** part of the current corresponds to $I_0(t)$; the **Imax2** part of the current to the first harmonic of $I_2(t)$; and the **Imax3** part of the current to the third harmonic of $I_2(t)$. Plotting of the currents for the considered and numerical methods are done as follows:

Isum=Imax+Imax2+Imax3;

Iapprox=Plot[{Part[Isum,1],Part[Isum,2]},{t,7/Ff,(7+1)/Ff},DisplayFunction
->Identity];

Iexact=Plot[Evaluate[{y[t],x[t]}/.sol],{t,7/Ff,8/Ff},DisplayFunction->Identity];

Show[{Iexact,Iapprox},DisplayFunction->$DisplayFunction];

Figure 4.4 presents the currents in the load calculated on the basis of the considered and numerical methods.

For the stability calculation, the expressions corresponding to instantaneous values of the power are defined in the cell:

Clear[t];

pt:=Part[Re[Simplify[((Ss1*(p-I*2*ω))/.p->I*2*ω)(Cos[2*ω*t]+

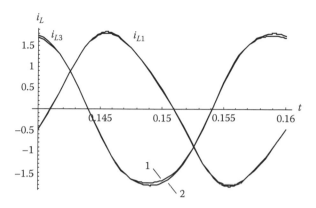

FIGURE 4.4
Currents i_{L1} and i_{L3} calculated on the basis of the considered (1) and numerical (2) methods (i_{I1} and i_{L3} in amperes, time t in seconds).

I*Sin[2*ω*t])]]*ki*ku+pcon,1,1];

pt2:=2*eqd2*Re[Simplify[((Simplify[Gd2*(p-I*2*ω)])/.{p->I*2*ω,φ->eqφ})

(Cos[2*ω*t]+I*Sin[2*ω*t])]]*ki*ku;

Using these expressions, we calculate the steady-state process stability by (4.31):

$$H4 = \begin{pmatrix} 1 & 0 \\ 0 & 1 \end{pmatrix};$$

kMax=100;

t0=10/Ff;

kpt2:=ku*ki*ky^2*(D1+1)*kd/((1+d0*ky*kd*(pt+pt2))^2);

For[k=1,k<=kMax, {t=t0+k/Ff/kMax;

H4=H4.MatrixExp[Linv.(-A11-kpt2*E11.Transpose[kc.Uri].kc)/Ff/kMax]};k++];

Eigenvalues[H4]

As a result we obtain

$$\{3.98072 \times 10^{-13}, 1.88012 \times 10^{-23}\}$$

Since the eigenvalues are less than unity the system is stable.

For an estimation of the given values, the stability calculation is realized on the basis of the data obtained by a numerical method:

$$H3 = \begin{pmatrix} 1 & 0 \\ 0 & 1 \end{pmatrix};$$

t0=6/Ff;

kMax=100;

kpsol:=Re[Part[ku*ki*ky^2*(D1+1)*kd/((1+ky*kd*Evaluate[pxy/.sol])^2),1]];

For[k=1,k<+kMax, {t=t0+k/Ff/kMax;

H3=H3.MatrixExp[Linv.(-A11-kpsol*E11.Transpose[kc.Urin].kc)/Ff/kMax]};k++];

Eigenvalues[H3]

As a result we have

$$\{4.15895 \times 10^{-13}, 1.54423 \times 10^{-23}\}$$

Comparing the calculated results, one can conclude that the eigenvalues differ slightly.

4.2 Characteristics of the Noncompensated DC Motor

4.2.1 Static Characteristics of the Noncompensated DC Motor

Noncompensated DC motors are widely used in DC drives in view of their more simple construction, high efficiency, and low cost. In particular, the exploitation of such motors differs from the exploitation of compensated DC motors. The difference results in the weakening of the resulting magnetic field, which changes the working characteristics of the motor (Korotyeyev and Klytta, 2005; Korotyeyev and Klytta, 2006). Consider a mathematical model of a noncompensated DC motor, the equivalent circuits of which are shown in Figure 4.5.

The differential equations describing the processes in a noncompensated DC motor, and taking into account the dependence of the magnetic field versus the current, have the form

$$L_A \frac{di_A(t)}{dt} = U_A - R_A i_A(t) - k\Phi(i_A(t))\omega(t);$$

$$J \frac{d\omega(t)}{dt} = k\Phi(i_A(t))i_A(t) - M_L.$$

(4.32)

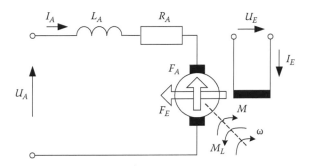

FIGURE 4.5
Equivalent circuits of a noncompensated DC motor. (Data from Korotyeyev I. Ye. and Klytta M., 2005. With permission.)

Assume that the dependence of the magnetic field versus the armature current about an operating mode is linear:

$$\Phi(I_A) = \Phi_N + \frac{\Delta\Phi_N}{I_N}(I_N - I_A) = a - bI_A, \tag{4.33}$$

where R_A and L_A are the resistance and inductance of an armature, respectively; $i_A(t)$ is the armature current; M_L is the load torque; k is the machine constant; $\omega(t)$ is the motor speed; J is the moment of inertia; Φ_N is the nominal flux; $\Delta\Phi_N = \Phi_{EN} - \Phi_N$; Φ_{EN} is the no-load flux; I_N is the nominal armature current; and a and b are constants.

A steady-state process is determined by equating to zero the right parts of the set (4.32):

$$U_A - R_A I_A - k\Phi(I_A)\omega = 0,$$
$$k\Phi(I_A)I_A - M_L = 0. \tag{4.34}$$

Static characteristics of the noncompensated DC motor are analyzed on the basis of the motor, with parameters $P_N = 7.6$ kW; $U_{AN}/I_{AN} = 420$ V/20.3 A; $n_N = 1950$ 1/min; $U_{EN}/I_{EN} = 210$ V/2.4 A; $R_A = 2.15\ \Omega$; $L_A = 13.4$ mH; $J = 23.7 \cdot 10^{-3}\ kgm^2$; $k = 251.1$; $\Phi_N = 7.3 \cdot 10^{-3}$ Vs; and $M_N = 37.2$ Nm.

Figure 4.6 presents the experimental magnetization curve of this motor.

Weakening of the resulting magnetic field in the noncompensated DC motor is shown in Figure 4.7, where the curve (2) corresponds to the real characteristics, whereas the curve (1) is calculated using (4.33).

The real characteristics of the influence of armature reaction for the different excitation currents are presented in Figure 4.8.

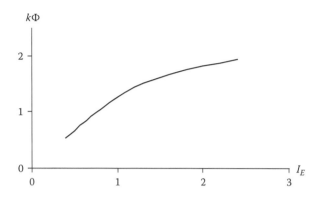

FIGURE 4.6
Magnetization characteristic of the analyzed motor ($k\Phi$ in volt-seconds, I_E in amperes). (Data from Korotyeyev I. Ye. and Klytta M., 2006. With permission.)

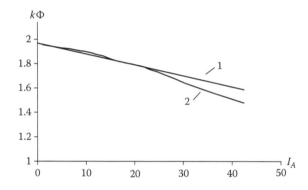

FIGURE 4.7
Field weakening for $I_E = I_{EN}$: computed curve (1) and measured characteristic (2) ($k\Phi$ in volt-seconds, I_E in amperes). (Data from Korotyeyev I. Ye. and Klytta M., 2005. With permission.)

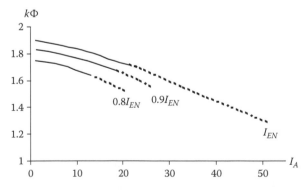

FIGURE 4.8
Measured field characteristics of the analyzed DC motor for the various excitation currents ($k\Phi$ in volt-seconds, I_E in amperes). (Data from Korotyeyev I. Ye. and Klytta M., 2006. With permission.)

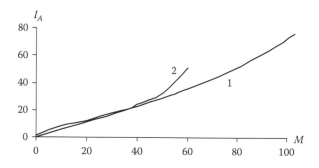

FIGURE 4.9

Armature current as a function of load torque for the nominal supply (U_{AN}, U_{EN}): Computed curve (1) and measured characteristic (2) (I_E in amperes, M in newton-meters). (Data from Korotyeyev I.Ye. and Klytta M., 2006a. With permission.)

In Figure 4.9 the theoretical (1) and real (2) curves of the armature current as a function of load torque are presented. From this figure one can see that, in the view of stronger weakening of the field, the armature current increases in comparison with the theoretical dependence.

Using the mathematical model, we plot the static characteristics of the noncompensated DC motor. In the cell, the values of the parameters are inputted:

$$Ra=2.15;$$

$$La=13.4*10^\wedge(-3);$$

$$\Phi n=7.3*10^\wedge(-3);$$

$$nN=1950;$$

$$Mn=37.2;$$

$$J1=23.7*10^\wedge(-3);$$

$$Inom=20.3;$$

In the next cell, the extrinsic parameters of the motor are calculated:

$$\Delta\Phi0=0.1;$$

$$\Delta\Phi=\Delta\Phi0*\Phi n;$$

$$a1=k*(\Phi n+\Delta\Phi);$$

$$b1=k*\Delta\Phi/Inom;$$

$$k\Phi=a1-b1*Inom;$$

$$kn=nN/\omega n;$$

$$msk=Mn/(\omega n^2);$$

$$k=Mn/Inom/\Phi n;$$

$$\omega n=N[2*Pi*nN/60];$$

The coefficient **kn** corresponds to the conversion of the rotational speed in the velocity.

In the next cell, equations that are necessary for plotting the torque-current characteristic are defined:

$$Un=420;$$

$$Imax=a1/2/b1$$

$$k\Phi1=0.8;$$

$$\omega s:=(Un-Ra*Ix)/(a1-b1*Ix);$$

$$Mx:=(a1-b1*Iy)*Iy;$$

$$Mx3:=(a1-b1*Iy2)*Iy2;$$

$$Mx5:=(a1-b1*Iy4)*Iy4;$$

$$Mx6:=(a1*k\Phi1-b1*Iy5)*Iy5;$$

$$Iy:=(Un-a1*nr/kn)/(Ra-nr/kn*b1);$$

$$Iy2:=(Un/2-a1*nr/kn)/(Ra-nr/kn*b1);$$

$$Iy4:=(Un/4-a1*nr/kn)/(Ra-nr/kn*b1);$$

$$Iy5:=(Un-a1*nr/kn*k\Phi1)/(Ra-nr/kn*b1);$$

In this cell, currents **Iy, Iy2, Iy4,** and **Iy5** are obtained from the first equation of the set (4.34) for different voltages **Un** and under weakening of the field defined by the **kΦ1** factor; **Mx, Mx3, Mx5,** and **Mx6** are obtained from the second equation of the set (4.34) for different currents **Iy, Iy2, Iy4,** and **Iy5**. It is also assumed that the expression **a1–b1*Iy** corresponds to $k\Phi(I_A)$. That means that **a1=a*k** and **b1=b*k**.

From the equations, the current value **Imax** corresponding to the moment maximum is determined. Then this value is used in plotting part of the dependence $M = f(I)$:

MIplot1=Plot[Mx,{Iy,0,Imax},AxesLabel->{"I","M"},DisplayFunction-> Identity];

MIplot2=Plot[Mx,{Iy,Imax,140},AxesLabel->{"I","M"},DisplayFunction-> Identity];

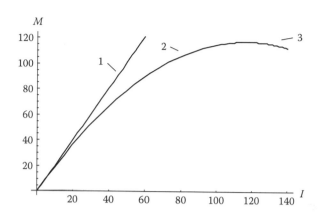

FIGURE 4.10
Curving of the torque characteristic of the noncompensated DC motor (*M* in newton-meters, *I* in amperes). (Data from Korotyeyev I.Yc. and Klytta M., 2006a. With permission.)

MIplot3=Plot[Iy*a1,{Iy,0,60},PlotStyle->{Thickness[0.004]},

AxesLabel->{"I","M"},DisplayFunction->Identity];

The torque-current characteristic shown in Figure 4.10 is plotted with the use of the function

Show[{MIplot1,MIplot2,MIplot3},DisplayFunction->$DisplayFunction];

The straight line (1) corresponds to the condition $M \sim I_A$, and the curve (2) corresponds to the condition $M \sim \Phi(I_A)I_A$. The part (3) of the characteristic corresponds to an unstable region of the work. The maximum value determines the critical torque as in the case of an asynchronous motor.

The torque-speed characteristics of the noncompensated DC motor are plotted with the use of the function

**Plot[{Mx,Mx3,Mx5,Mx6},{nn,0,2674},AxesLabel"ω","M"},PlotRange->
{0,150}];**

These characteristics are presented in Figure 4.11. From the figure one sees that the characteristics of the noncompensated DC motor and an asynchronous motor are similar.

The real characteristics of the motor for various voltages that are less than nominal is presented in Figure 4.12, which practically coincide with the calculated ones. Differences become apparent for voltages close to nominal one.

The excitation weakening causes the critical torque to decrease. The real characteristics of the torque versus speed for $I_E = 0.8\ I_{EN}$ are shown in Figure 4.13.

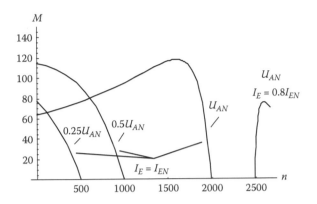

FIGURE 4.11
Calculated M/n characteristics of the analyzed motor (M in newton-meters, n in rpm).

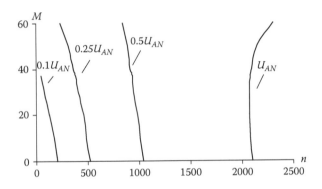

FIGURE 4.12
Real M/n characteristics of the analyzed motor for $I_E = I_{EN}$ and various armature voltages (M in newton-meters, n in rpm). (Data from Korotyeyev I. Ye. and Klytta M., 2006a. With permission.)

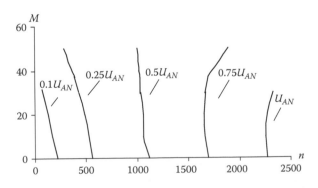

FIGURE 4.13
Real M/n characteristics of the analyzed motor for $I_E = 0.8\,I_{EN}$ and various armature voltages (M in newton-meters, n in rpm). (Data from Korotyeyev I. Ye. and Klytta M., 2006a. With permission.)

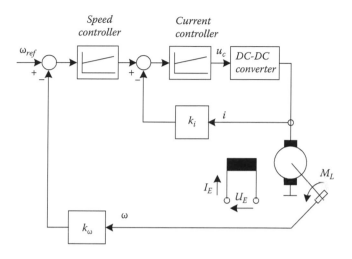

FIGURE 4.14
Block diagram of the DC drive control system with a noncompensated DC motor. (Data from Korotyeyev I. Ye. and Klytta M., 2006b.)

4.2.2 Analysis of Electrical Drive with Noncompensated DC Motor

Let us consider the starting characteristics in a DC drive with a noncompensated DC motor for various load torques (Korotyeyev and Klytta, 2006b). The typical system for the speed control of the DC drive with the additional current loop is shown in Figure 4.14.

The parameters of the PI controllers obtained by an empirical method are the gains $k_s = 4$, $k_c = 0.7$, the integration time constants $T_s = 99.8$ ms, $T_c = 0.952$ s. Other parameters of the control system are the gains $k_i = 0.5$, $k_\omega = 5{,}4$ 10^{-3}, and the reference speed $\omega_{ref} = 5$. Further, we compare the start characteristics for two load torques:

Constant load torque

$$M_L = \text{const},$$

Quadratic load torque

$$M_L = m_\omega \omega^2. \tag{4.35}$$

The coefficient $m_\omega = 0.932 \cdot 10^{-3}$ is determined from the nominal point:

$$m_\omega = M_N / \omega_N^2.$$

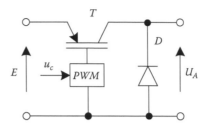

FIGURE 4.15
Buck converter as controlled supply source of a DC motor. (Data from Korotyeyev I. Ye. and Klytta M., 2006b.)

The DC motor is supplied by the Buck converter shown in Figure 4.15 with the PWM control of the output voltage.

The switching frequency of the converter (IGBT technology) equals 10 kHz, and the PWM control signal has a "sawtooth" form, with the magnitude of $U_g = 5$ V. The converter is supplied by a 520 V DC voltage source.

The parameters of the control system are presented in the cell:

$$Un=520;$$

$$tk=4.0;$$

$$KyI=0.5;$$

$$Tel=1.05;$$

$$Ky=4.0;$$

$$Kd=0.0054;$$

$$Kel=0.7;$$

$$Tem=10.02;$$

In this cell, **tk** defines the duration of the calculation of a transient behavior, **KyI** defines the gain of the current controller, **Tel** defines the inverse value of the time constant of the current controller, **Kd** denotes k_ω, **Ky** defines the gain of the speed controller, and **Tem** defines the inverse value of the time constant of the speed controller. The current and speed controller are defined as follows:

$$KyI + \frac{Tel}{s},$$

$$Ky + \frac{Tem}{s}.$$

The calculation of the speed shows that the ripple of the speed is small for the high-modulation frequency. This allows the use of the average state-space method (Middlebrook and Ćuk, 1976) for solving differential equations that describe the processes in the system shown in Figure 4.14.

In the next cell, the equations for the noncompensated DC motor with constant load torque are presented. The solution to these equations is determined with the use of the **ND Solve** [] function.

$$Ieq5:=(yi[t]*Un/Ug-(a1-b1*it[t])*\omega t[t])/La-Ra*it[t]/La;$$

$$\Omega eq5:=(a1-b1*it[t])*it[t]/J1-Mn/J1;$$

$$Zeq5:=Tem*(Uref-Kd*\omega t[t])-Ky*Kd*\ \Omega eq5;$$

$$Yeq5:=Tel*(yz[t]-KyI*it[t])+Kel*(Zeq5-KyI*Ieq5);$$

$$sol6=NDSolve[\{it'[t]==Ieq5,\omega t'[t]==\Omega eq5,yi'[t]==Yeq5,yz'[t]==Zeq5,$$

$$it[0]==0,\omega t[0]==0,yi[0]==Kel*Ky*Uref,yz[0]==Ky*Uref\},$$
$$\{it,\omega t,yi,yz\},\{t,0,tk\}];$$

In this cell, **Uref** denotes ω_{ref}, **Ieq5** and **Ωeq5** define the two parts of (4.32); **Zeq5** defines an equation of the speed controller; and **Yeq5** defines an equation of the current controller.

The starting characteristics of the current and speed are plotted as follows:

$$Pusr\Omega const=Plot[Evaluate[\omega t[t]/.sol6],\{t,0,tk\},AxesLabel->$$

$$\{"t","\omega"\},PlotRange->All];$$

$$PusrIconst=Plot[Evaluate[it[t]/.sol6],\{t,0,tk\},AxesLabel->$$

$$\{"t","i"\},PlotRange->All];$$

When the load torque is dependent on the square of speed (4.35), the equations have the form

$$\Omega eq6:=(a1-b1*it[t])*it[t]/J1-msk*\omega t[t]*\ \omega t[t]/J1;$$

$$Zeq6:=Tem*(Uref-Kd*\omega t[t])-Ky*Kd*\ \Omega eq6;$$

$$Yeq6:=Tel*(yz[t]-KyI*it[t])+Kel*(Zeq6-KyI*Ieq5);$$

$$sol7=NDSolve[\{it'[t]==Ieq5,\ \omega t'[t]==\ \Omega eq6,yi'[t]==Yeq6,yz'[t]==Zeq6,$$

$$it[0]==0,\omega t[0]==0,yi[0]==Kel*Ky*Uref,yz[0]==Ky*Uref\},\{it,\omega t,yi,yz\},\{t,0,tk\}];$$

The starting characteristics are plotted in the same way:

PusrΩ=Plot[Evaluate[ωt[t]/.sol7],{t,0,tk},AxesLabel->{"t","ω"},

PlotRange->All];

PusrI1=Plot[Evaluate[it[t]/.sol7],{t,0,tk},AxesLabel->{"t","i"},

PlotRange->All,DisplayFunction->Identity];

Let us calculate the starting characteristics of a compensated DC motor and compare them with the ones obtained earlier. The mathematical model of the compensated DC motor follows from (4.32):

$$L_A \frac{di_A(t)}{dt} = U_A - R_A i_A(t) - k\Phi_N \omega(t);$$

$$J \frac{d\omega(t)}{dt} = k\Phi_N i_A(t) - M_L.$$

The parameters for the compensated DC motor are the same as chosen for the noncompensated DC motor. The equations for the constant load torque are

Ieq7:=(yi[t]*Un/Ug–kΦ*ωt[t])/La–Ra*it[t]/La;

Ωeq7:=kΦ*it[t]/J1–Mn/J1;

Zeq7:=Tem*(Uref–Kd*ωt[t])–Ky*Kd* Ωeq7;

Yeq6:=Tel*(yz[t]–KyI*it[t])+Kel*(Zeq7–KyI*Ieq7);

sol8=NDSolve[{it′[t]==Ieq7,ωt′[t]== Ωeq7,yi′[t]==Yeq7,yz′[t]==Zeq7,

it[0]==0,ωt[0]==0,yi[0]==Kel*Ky*Uref,yz[0]==Ky*Uref},{it,ωt,yi,yz},{t,0,tk}];

The starting characteristics are plotted with the use of the functions

PusrΩconstSkom=Plot[Evaluate[ωt[t]/.sol8],{t,0,tk},AxesLabel->{"t","ω"},

PlotRange->All];

PusrIconstSkom1=Plot[Evaluate[it[t]/.sol8],{t,0,tk},AxesLabel->{"t","i"},

PlotRange->All,DisplayFunction->Identity];

The equations describing the processes in the compensated DC motor for the load torque (4.35) have the form

Ieq8:=(yi[t]*Un/Ug–kΦ*ωt[t])/La–Ra*it[t]/La;

Ωeq8:=kΦ*it[t]/J1–msk*ωt[t]*ωt[t]/J1;

Zeq8:=Tem*(Uref-Kd*ωt[t])-Ky*Kd* Ωeq8;

Yeq8:=Tel*(yz[t]-KyI*it[t])+Kel*(Zeq8-KyI*Ieq7);

sol9=NDSolve[{it'[t]==Ieq8, ωt'[t]== Ωeq8,yi'[t]==Yeq8,yz'[t]==Zeq8,

it[0]==0, ωt[0]==0,yi[0]==Kel*Ky*Uref,yz[0]==Ky*Uref},
{it, ωt,yi,yz},{t,0,tk}]];

The starting characteristics are prepared for plotting as follows:

PusrΩvarSkom=Plot[Evaluate[ωt[t]/.sol9],{t,0,tk},AxesLabel->{"t","ω"},

PlotRange->All];

PusrIvarSkom1=Plot[Evaluate[it[t]/.sol9],{t,0,tk},AxesLabel->{"t","i"},

PlotRange->All,DisplayFunction->Identity];

All starting characteristics of the armature current are plotted with the help of the function

Show[PusrIconst1,PusrI1,PusrIconstSkom1,PusrIvarSkom1,

DisplayFunction->$DisplayFunction];

and are shown in Figure 4.16. In this figure there are presented transients: noncompensated motor with constant load torque $M_L = M_N$ (1); noncompensated motor with load torque proportional to the square of speed (2); compensated DC motor with constant load torque (3); compensated DC motor with load torque proportional to the square of speed (4).

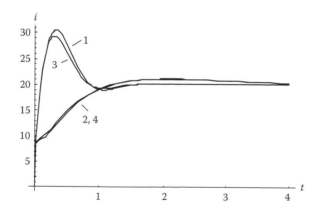

FIGURE 4.16
Armature current starting characteristics in DC drive with compensated and noncompensated DC motors (*i* in amperes, time *t* in seconds). (Data from Korotyeyev I. Ye. and Klytta M., 2006b.)

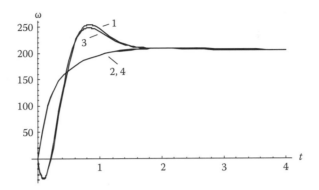

FIGURE 4.17
Motor speed starting characteristics in DC drive with compensated and noncompensated DC motors (ω in radian/second, time t in seconds). (Data from Korotyeyev I. Ye. and Klytta M., 2006b.)

The starting characteristics of the speed for different load toques and two motor types are displayed with the help of the function

Show[PusrΩconst,PusrΩ,PusrΩconstSkom,PusrΩvarSkom];

and are shown in Figure 4.17.

To reduce the transient time, the parameters of PI regulators are optimized by the use of the module and symmetry criterions. For the current regulator, one obtains the parameters

$$k_c = 2.56; \quad T_c = 6.17 \text{ ms,}$$

and for the speed regulator,

$$k_s = 35.2; \quad T_s = 50.0 \text{ ms.}$$

The parameters of this case are inputted into the cell:

$$\text{Un=520;}$$

$$\text{tk=1.0;}$$

$$\text{KyI=0.5;}$$

$$\text{Kd=0.005;}$$

$$\text{Kd=0.0056;}$$

$$\text{Tel=162.0;}$$

$$\text{Ky=35.2;}$$

$$\text{Kel=2.56;}$$

$$\text{Tem=176.0;}$$

The equations for calculating the starting characteristics are written as follows:

For the DC drive with the noncompensated DC motor and constant torque,

$$Ieq5:=((UnitStep[yi[t]-Ug]+yi[t]/Ug*UnitStep[-yi[t]+Ug])*$$

$$Un-(a1-b1*it[t])*\omega t[t])/La-Ra*it[t]/La;$$

$$\Omega eq5:=(a1-b1*it[t])*it[t]/J1-Mn/J1;$$

$$Zeq5:=Tem*(Uref-Kd*\omega t[t])-Ky*Kd*\Omega eq5;$$

$$Yeq5:=Tel*(yz[t]-KyI*it[t])+Kel*(Zeq5-KyI*Ieq5);$$

$$sol6=NDSolve[\{it'[t]==Ieq5, \omega t'[t]== \Omega eq5,yi'[t]==Yeq5,yz'[t]==Zeq5,$$

$$it[0]==0, \omega t[0]==0,yi[0]==Kel*Ky*Uref,yz[0]==Ky*Uref\},$$
$$\{it, \omega t,yi,yz\},\{t,0,tk\}];$$

(the duty factor clipping is carried out by the **UnitStep[]** function).

For the DC drive with the noncompensated DC motor and the load torque proportional to the square of speed,

$$\Omega eq6:=(a1-b1*it[t])*it[t]/J1-msk*\omega t[t]* \omega t[t]/J1;$$

$$Zeq6:=Tem*(Uref-Kd*\omega t[t])-Ky*Kd* \Omega eq6;$$

$$Yeq6:=Tel*(yz[t]-KyI*it[t])+Kel*(Zeq6-KyI*Ieq5);$$

$$sol7=NDSolve[\{it'[t]==Ieq5, \omega t'[t]== \Omega eq6,yi'[t]==Yeq6,yz'[t]==Zeq6,$$

$$it[0]==0, \omega t[0]==0,yi[0]==Kel*Ky*Uref,yz[0]==Ky*Uref\},$$
$$\{it, \omega t,yi,yz\},\{t,0,tk\}];$$

For the DC drive with the compensated DC motor and the constant torque,

$$Ieq7:=((UnitStep[yi[t]-Ug]+yi[t]/Ug*UnitStep[-yi[t]+Ug])*Un-$$

$$k\Phi* \omega t[t])/La-Ra*it[t]/La;$$

$$\Omega eq7:=k\Phi*it[t]/J1-Mn/J1;$$

$$Zeq7:=Tem*(Uref-Kd*\omega t[t])-Ky*Kd* \Omega eq7;$$

$$Yeq7:=Tel*(yz[t]-KyI*it[t])+Kel*(Zeq7-KyI*Ieq7);$$

$$sol8=NDSolve[\{it'[t]==Ieq7, \omega t'[t]== \Omega eq7,yi'[t]==Yeq7,yz'[t]==Zeq7,$$

$$it[0]==0, \omega t[0]==0,yi[0]==Kel*Ky*Uref,yz[0]==Ky*Uref\},$$
$$\{it, \omega t,yi,yz\},\{t,0,tk\}];$$

For the DC drive with the compensated DC motor and the load torque proportional to the square of speed,

$$Ieq8:=((UnitStep[yi[t]-Ug]+yi[t]/Ug*UnitStep[-yi[t]+Ug])*Un-$$

$$k\Phi*\omega t[t])/La-Ra*it[t]/La;$$

$$\Omega eq8:=k\Phi*it[t]/J1-msk*\omega t[t]*\omega t[t]/J1;$$

$$Zeq8:=Tem*(Uref-Kd*\omega t[t])-Ky*Kd*\Omega eq8;$$

$$Yeq8:=Tel*(yz[t]-KyI*it[t])+Kel*(Zeq8-KyI*Ieq8);$$

sol9=NDSolve[{it'[t]==Ieq8, ωt'[t]== Ωeq8,yi'[t]==Yeq8,yz'[t]==Zeq8,it[0]==0,ωt[0]==0,

$$yi[0]==Kel*Ky*Uref,yz[0]==Ky*Uref\},\{it, \omega t,yi,yz\},\{t,0,tk\}];$$

The starting characteristics are obtained in the same way and are presented in Figures 4.18 and 4.19.

From the figure one observes some speed overshoots. For overcoming them, we form the reference signal as follows:

$$\omega_{ref} = \begin{cases} \omega_r t/Tref, & t \le T_{ref}, \\ \omega_r, & t > T_{ref}; \end{cases}$$

The parameters of this signal are $T_{ref} = 0.24$ s and $\omega_r = 5$.

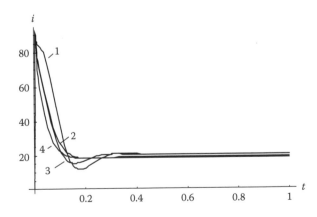

FIGURE 4.18
Armature current starting characteristics in DC drive with the parameters determined by the module and symmetry criterions (*i* in amperes, time *t* in seconds). (Data from Korotyeyev I. Ye. and Klytta M., 2006b.)

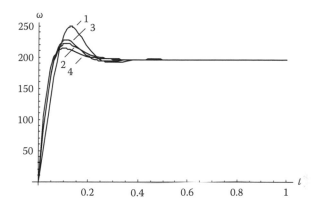

FIGURE 4.19
Motor speed starting characteristics in DC drive with the parameters determined by the module and symmetry criterions (ω in radian/second, time t in seconds). (Data from Korotyeyev I. Ye. and Klytta M., 2006b.)

The parameters for this case are presented in the cell:

Un=520;

tk=1.0;

KyI=0.5;

Kd=0.005;

Kd=0.0056;

Tel=162.0;

Ky=35.2;

Kel=2.56;

Tem=176.0;

Tref=0.24;

The equations for calculating the starting characteristics are written as follows:

For the DC drive with the noncompensated DC motor and the constant torque,

UrefT:=If[t>Tref,Uref,Uref*t/Tref];

Ieq5:=(yi[t]*Un/Ug-(a1-b1*it[t])* ωt[t])/La-Ra*it[t]/La;

$$\Omega eq5:=(a1-b1*it[t])*it[t]/J1-Mn/J1;$$

$$Zeq5:=Tem*(UrefT-Kd*\omega t[t])-Ky*Kd*\ \Omega eq5;$$

$$Yeq5:=Tel*(yz[t]-KyI*it[t])+Kel*(Zeq5-KyI*Ieq5);$$

$$sol6=NDSolve[\{it'[t]==Ieq5,\omega t'[t]==\ \Omega eq5,yi'[t]==Yeq5,yz'[t]==Zeq5,$$

$$it[0]==0,\omega t[0]==0,yi[0]==0,yz[0]==0\},\{it,\omega t,yi,yz\},\{t,0,tk\}];$$

(the representation of linear function of speed is realized by the $U_{ref}T$ function).

For the DC drive with the noncompensated DC motor and the load torque proportional to the square of speed,

$$\Omega eq6:=(a1-b1*it[t])*it[t]/J1-msk*\omega t[t]*\omega t[t]/J1;$$

$$Zeq6:=Tem*(UrefT-Kd*\omega t[t])-Ky*Kd*\ \Omega eq6;$$

$$Yeq6:=Tel*(yz[t]-KyI*it[t])+Kel*(Zeq6-KyI*Ieq5);$$

$$sol7=NDSolve[\{it'[t]==Ieq5,\omega t'[t]==\Omega eq6,yi'[t]==Yeq6,yz'[t]==Zeq6,$$

$$it[0]==0,\omega t[0]==0,yi[0]==0,yz[0]==0\},\{it,\omega t,yi,yz\},\{t,0,tk\}];$$

For the DC drive with the compensated DC motor and the constant torque,

$$Ieq7:=(yi[t]*Un/Ug-k\Phi*\ \omega t[t])/La-Ra*it[t]/La;$$

$$\Omega eq7:=k\Phi*it[t]/J1-Mn/J1;$$

$$Zeq7:=Tem*(UrefT-Kd*\omega t[t])-Ky*Kd*\ \Omega eq7;$$

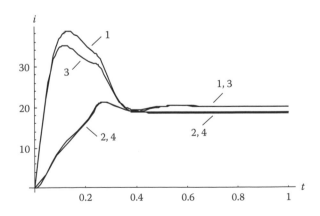

FIGURE 4.20
Armature current starting characteristics in DC drive using the set-point adjuster (i in amperes, time t in seconds). (Data from Korotyeyev I. Ye. and Klytta M., 2006b.)

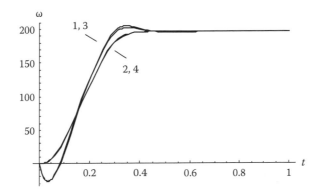

FIGURE 4.21
Starting transients of the motor speed in DC drive using the set-point adjuster (ω in radian/second, time t in seconds). (Data from Korotyeyev I. Ye. and Klytta M., 2006b.)

$$Yeq7:=Tel*(yz[t]–KyI*it[t])+Kel*(Zeq7–KyI*Ieq7);$$

$$sol8=NDSolve[\{it'[t]==Ieq7,\omega t'[t]==\Omega eq7,yi'[t]==Yeq7,yz'[t]==Zeq7,$$

$$it[0]==0,\omega t[0]==0,yi[0]==0,yz[0]==0\},\{it,\omega t,yi,yz\},\{t,0,tk\}];$$

For the DC drive with the compensated DC motor and the load torque proportional to the square of speed,

$$Ieq8:=(yi[t]*Un/Ug–k\Phi*\ \omega t[t])/La–Ra*it[t]/La;$$

$$\Omega eq8:=k\Phi*it[t]/J1–msk*\omega t[t]*\ \omega t[t]/J1;$$

$$Zeq8:=Tem*(UrefT-Kd*\omega t[t])–Ky*Kd*\ \Omega eq8;$$

$$Yeq8:=Tel*(yz[t]–KyI*it[t])+Kel*(Zeq8–KyI*Ieq8);$$

$$sol9=NDSolve[\{it'[t]==Ieq8,\omega t'[t]==\ \Omega eq8,yi'[t]==Yeq8,yz'[t]==Zeq8,$$

$$it[0]==0,\omega t[0]==0,yi[0]==0,yz[0]==0\},\{it,\omega t,yi,yz\},\{t,0,tk\}];$$

The current and speed starting characteristics are prepared and plotted in the same way. These characteristics are presented in Figures 4.20 and 4.21.

One can see that the overshooting of the armature current is smaller than in previous cases. It should be noted that the reduction of the armature currents during the transient process leads to an increase in the time of the transient process.

5

Modeling of Processes Using PSpice®

PSpice® standard is a computer program dedicated to process modeling in electrical circuits and, for the present, is a trademark of Cadence Company. The program packet is distributed under a few versions and names, mainly Microsim Design Lab and Orcad Family. The abbreviation SPICE means *Simulation Program with Integrated Circuit Emphasis*. Free or shareware versions of Spice are available. The common parts of all the versions are these modules:

- Simulation Manager
- Schematic editor
- PSpice AD

On the basis of the circuits considered in the previous chapters, in particular, how to use these programs to create and manage circuit drawings, set up and run simulations, and evaluate simulation test results will be shown. The results of the simulations will be compared with those obtained from the Mathematica® models.

5.1 Modeling of Processes in Linear Systems

The principles of operation of the PSpice schematic editor will be shown using the schematic diagram from Chapter 2, Figure 2.1. Let us draw the circuit presented in Figure 5.1. Start the schematic editor by double-clicking on the **Schematics** icon in the program group. An empty schematic page is displayed. In the beginning, it is recommended the design be named and saved by the **Save_As** command from the **File** dialog box.

5.1.1 Placing and Editing Parts

All the parts are marked by a name with a number, for example, L1 or V1. The proper part can be found in the **Part Browser**, and the dialog box is marked in Figure 5.1 as (1). Typing the name (L, V, etc.) and then pressing Enter or OK

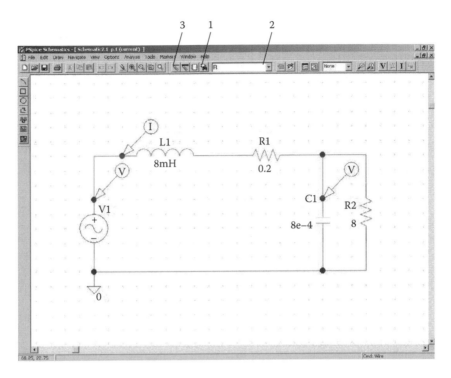

FIGURE 5.1
Exemplary schematics page.

enables selection of the part to be placed. The 10 most recent parts are stored in the list (2). We can move the selected symbol to its location and right-click to stop placing the parts. The part already placed can be flipped or rotated by pressing the Ctrl-F or Ctrl-R buttons. To connect parts with a wire, use the button **Draw Wire** (3) by right-clicking the mouse to begin, as well as to finish.

An **Agnd** (analog ground) or a **Gnd** component must be placed and connected to one node of our design. It is necessary to fulfill the electrical rules of the PSpice netlisting. (Muhammad H. Rashid, Hassan M Rashid 2005)[23]

5.1.2 Editing Part Attributes

All the schematic parts and symbols have associated attributes. They can be edited by double-clicking on the part, for example, the V1 source, as shown in Figure 5.2. The attribute dialog box is opened. It is necessary to fill the attribute empty fields.

In our example, the part values should be as follows:

R1 = 0.2 Ω
R2 = 8 Ω
L1 = 8 mH

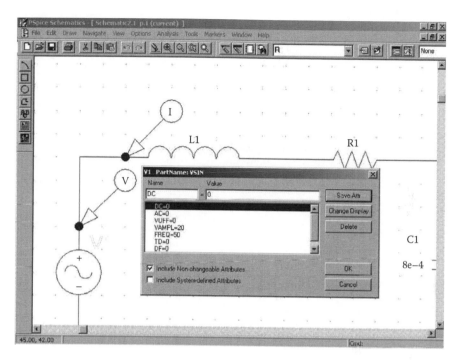

FIGURE 5.2
Part attribute dialog box.

C1 = 80 uF

V1 = Vsin {Vampl = 20 V; Freq = 50, Td = 0 Hz, phase = 0}

Attributes indicated with the '*' symbol are fixed and cannot be changed or deleted in the schematic editor. However, they can be globally modified in the **Symbol Editor**.

The **V** and **I markers** placed in our schematics determine values that we want to be automatically performed as the result of analysis.

5.1.3 Setting Up Analyses

Standard PSpice A/D analyses are as follows:

- **DC Sweep**—Currents and voltages of the steady-state response are calculated.

- **Bias Point Detail**—The bias point is automatically computed by **PSpice A/D**; selecting this item results in reporting the data in output files.

- **DC sensitivity**—This calculates the sensitivity of a node or component's voltage as a function of the bias point.

- **AC Sweep** (frequency response)—This calculates the small-signal response of the circuit to a combination of inputs. The sources are swept over a declared frequency range. Magnitudes and phases of the output values are calculated.

- **Noise Analysis**—This is performed with frequency response analysis. For every frequency specified in the analysis, the contribution of each noise generator in the circuit is transferred to an output node.

- **Transient Response**—The behavior of the circuit is observed over time as a response of time-varying parameters.

- **Fourier Components**—This can be performed with transient analysis. It calculates the Fourier components of selected signals.

5.2 Analyzing the Linear Circuits

5.2.1 Time-Domain Analysis

To analyze the circuit in our example in the time domain, it is necessary to open the **Analysis**, and then the **Setup** dialog boxes. As shown in Figure 5.3, first the **Bias point detail** and then the **Transient** must be selected. As a result, the proper dialog box, as shown in the figure, should be opened. In this box the parameter **Print Step** determines the time intervals for saving values in output files; the **Final Time** determines the duration of the analysis. The

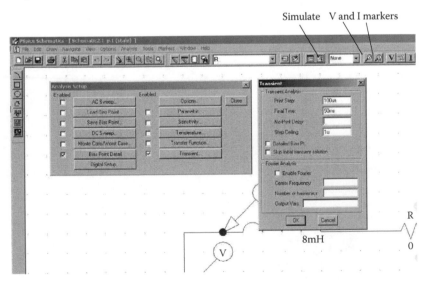

FIGURE 5.3
Setting up the transient analysis.

parameter **No-Print Delay** determines the moment to start saving the simulation results—the previous values will be computed but not saved in the output files. The **Step Ceiling** determines the maximum allowable time step size for a computing algorithm. Parameters for our analysis should be set as

Print Step = 100 μs

Final Time = 50 ms

No-Print Delay = 0 or empty

Step Ceiling = 1 μs

When our design is finished, we can start to simulate it. Pressing the **Simulate** or **F11** buttons will start the specified analyses. The **PSpice A/D** module starts to compute the processes in the circuit and save them. First, our design is checked for errors. In the case when the circuit or a parameter is incorrect, the error is described in an output file. The output file can be opened from the **Analysis** dialog box, under the **Examine Output** option. For example, if, in the R1 value, "0,1" is written instead of "0.1", the proper fragment of the output is shown as follows:

```
* Schematics Netlist *
R_R2        0 $N_0001 8
C_C1        $N_0001 0 8e-4
V_V1        $N_0002 0 DC 0 AC 0
+SIN 0 20 50 0 0 0
L_L1        $N_0002 $N_0003 8mH
R_R1        $N_0003 $N_0001 0,2
-----------------------------$
ERROR -- Value may not be 0
**** RESUMING Schematic2.1.cir ****
```

The error "Value may not be 0" has been found during netlisting of the schematic and pointed out in the output file in the line before the sign $. It is necessary to correct the error in the proper **Schematics** field and start the simulation again. Correctness of the circuit is checked again, then the bias point for the transient analysis is calculated. Next, the transient analysis starts up, and when it is finished, the **Probe** window appears. The results of the simulation are written in output files with the same name as the circuit but with the .dat and .out extensions. This means that the **Probe** can be started next time without simulation of the circuit.

When the analysis is finished and the **Probe** window is opened, as shown in Figure 5.4, the easiest way to display the traces is to label them in the **Schematics** with proper voltage or current markers, shown in Figure 5.1.

It should be noted that markers can be inserted also after analyses. Marked traces are displayed automatically. The other traces that were not marked in

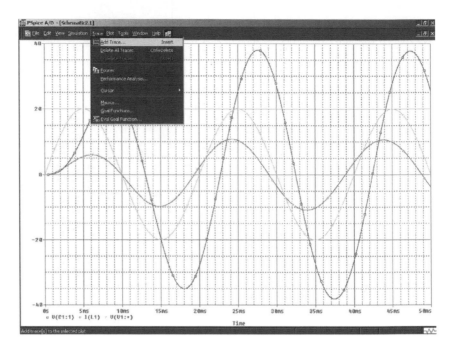

FIGURE 5.4
Probe window with marked traces.

the **Schematics** can be selected from the **Trace** menu by the **Add** option (or just by pressing the Insert key). A list of all available traces will be displayed. We can choose the desired trace from the list and click OK.

The easy way to copy the results of our simulation to other Windows programs is to use the **Copy to clipboard** option from the **Window** dialog box. All of the figures presenting simulation test results were obtained in this way.

The range of the time axis is the same as that used in the simulation profile. The means that the beginning moment is equal to 0 or that the **No-Print Delay** value and the end is equal to the **Final Time** of the analysis. The time range, or even the axis variable of displayed results, can be changed by double-clicking on the time axis. The results of our simulation are shown in Figure 5.5 and are similar to those from Chapter 2, Figure 2.2. They are obtained by setting the time range from 0 to 20 ms.

If we need to obtain a proper phase-plane portrait, we can do it by choosing for the **x axis variable** the I(L1) value, and for the **trace** the V(C1:1) value. The trajectory is presented in Figure 5.6 and looks similar to the one from the **Mathematica** model, shown in Chapter 2, Figure 2.3.

Moreover, the **Probe** allows presentation of many other time-dependent values. This is possible using the **Analog Operators and Functions** from the right of the **Add Trace** menu. For example, if we need to observe an RMS value of the input voltage, we can perform it by typing: RMS(V(V1:+)). To obtain the

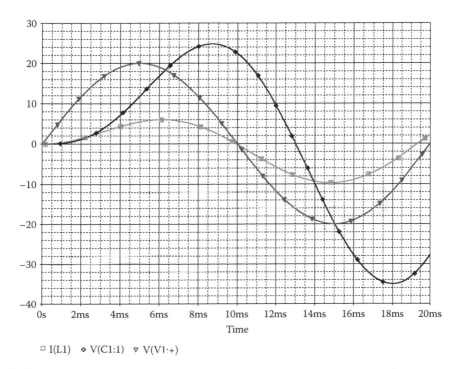

FIGURE 5.5
Time waveforms of analyzed circuit: input voltage V(V1+), inductor current I(L1), and capacitor voltage V(C1:1).

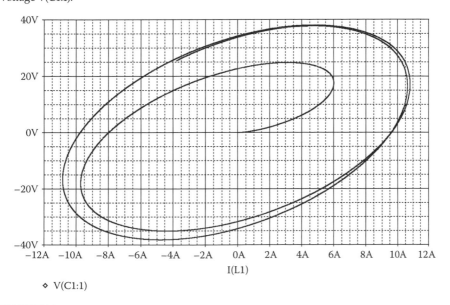

FIGURE 5.6
The phase-plane trajectory V(C1:1) versus I(L1).

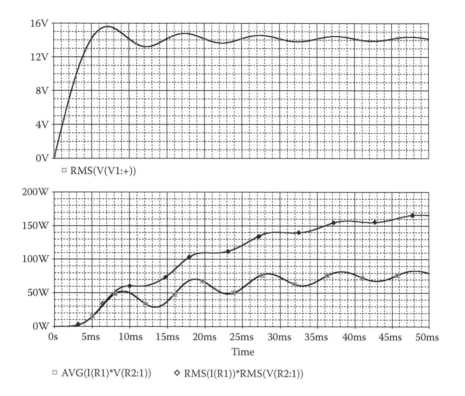

FIGURE 5.7
Time waveforms of the RMS value of the input voltage (upper window), active and apparent power of the load (lower window).

active power dissipated in the load of our circuit, type AVG(I(R1)*V(R2:1)); and for apparent power, type RMS(I(R1))* RMS(V(R2:1)). Actual waveforms are presented in Figure 5.7.

5.2.2 AC Sweep Analysis

AC sweep analysis is a linear analysis in the frequency domain. It presents the frequency response of a circuit over a user-defined frequency range of AC sources. Let us analyze our linear circuit. It is necessary to change the **Vsin** voltage source to a **VAC** component that is proper for this simulation profile, as shown in Figure 5.8. Instead of time-domain analysis we must choose **AC Sweep and Noise Analysis**. Set the parameters of the simulation as follows:

- AC Sweep Type: Linear
- Total Points: 101

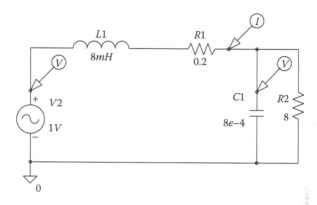

FIGURE 5.8
Schematic diagram for **AC Sweep** analysis.

- Start Frequency: 1 Hz
- End Frequency: 200 Hz

In this case, our circuit will be linearly examined for frequency response from 1 to 200 Hz, and 101 points of characteristics will be saved in the output file. The results of the simulation are presented in Figure 5.9.

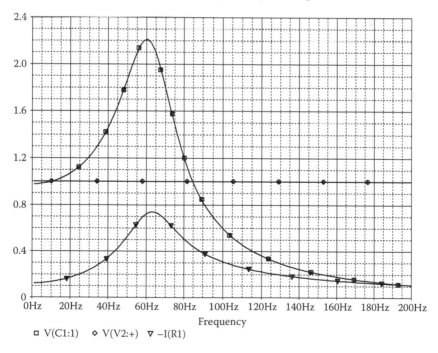

FIGURE 5.9
Results of the **AC Sweep** and **Noise Analysis**.

If the selected values are like those from the figure, we can observe the constant value of the input voltage 1V and the frequency-dependent load voltage and current. As can be seen, magnitudes of the AC sources are constant, but their frequencies vary.

5.3 Modeling of Nonstationery Circuits

5.3.1 Transient Analysis of a Thyristor Rectifier

Figure 5.10 presents a PSpice model of the thyristor (SCR) rectifier presented in Chapter 2. This is a direct AC/DC converter with a thyristor-diode bridge with an RL load. The output voltage depends on the delay between the moment of zero crossing of the AC voltage and the rising edge of the gate impulse. This control function is realized by a TD parameter in a **V2** pulse voltage. Its parameters are V1 = 0, V2 = 10, TD = 2 ms, TR = 100 μs, TF = 100 μs, PW = 1 ms, PER = 10 ms. The gate pulse is the same for both thyristors, and its frequency is twice the input voltages. In each moment, there could be only one thyristor that conducts a load current. That is, the ignited thyristor will be the one with the highest anode potential.

The valves X1 and X2 are described by 2N2579–600 V 25 A type, which model is defined by the manufacturer. The AC source is modeled by the VSin sinusoidal voltage with internal parameters VAMPL = 310 V, FREQ = 50 Hz, TD = 0, Phase = 0. For convenient observation of the load voltage, the negative DC terminal is grounded.

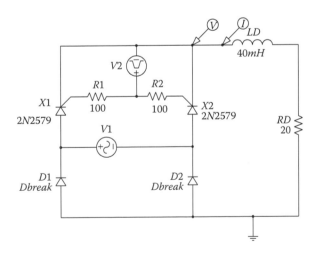

FIGURE 5.10
A model of the phase-controlled rectifier.

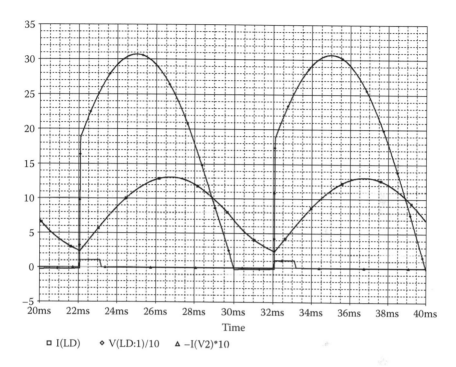

FIGURE 5.11
Load current, load voltage, and gate impulses.

Exemplary load current, voltage and gate pulses are shown in Figure 5.11. For better clearance, the voltage and impulses were scaled 10 times. Because of the relatively small value of the time constant L/R of the load, the steady state of the circuit was reached in less than one input AC voltage period.

One can see that the curve of the load current presented in Figure 2.7 is the same as the one presented in Figure 5.11.

5.3.2 Boost Converter—Transient Simulation

The open-loop system with the Boost converter is presented in Figure 5.12. The switching elements in the circuit are the diode—a **Dbreak** diode—and an **S** switch.

A **Dbreak** model is the built-in standard silicon diode model described in the **Breakout.slb** and **Breakout.plb** model libraries. The **S** switch, also described in the **Breakout** library, must be previously defined using the **Part Attribute** dialog box. Actually, this component is represented as a voltage-controlled resistance. Parameters to be defined are the on- and off-state

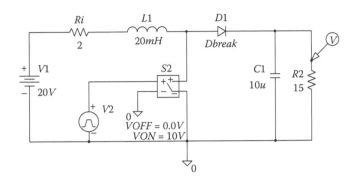

FIGURE 5.12
Schematic diagram of the Boost converter.

resistances and the thresholds v-on and v-off voltages corresponding to them. In our case, the parameters are

- Roff = 1e6 Ω
- Ron = 2 Ω
- Von = 10 V
- Voff = 0 V

As the generator of cyclic pulses controlling the switch, the **Vpulse** trapezoidal voltage source is used. Its parameters should be set as

- DC = 0—the voltage for **Bias Point** calculation
- AC = 0—AC magnitude, used for **AC sweep** analysis only
- V1 = 0—initial voltage value
- V2 = 10 V—pulsed voltage value
- TD = 0 s—delay time
- TR = 100 ns—rise time
- TF = 100 ns—fall time
- PW = 0.000469—pulse width
- PER = 1 ms—period

The description of the **Vpulse** component contains some **Simulationonly** parameters. It indicates symbols to be used for a simulation but not for a board layout. In our case, the field can be left empty.

The circuit is analyzed for transients for 8 ms of time. The transient states of the input current and output voltage are shown in Figures 5.13 and 5.14.

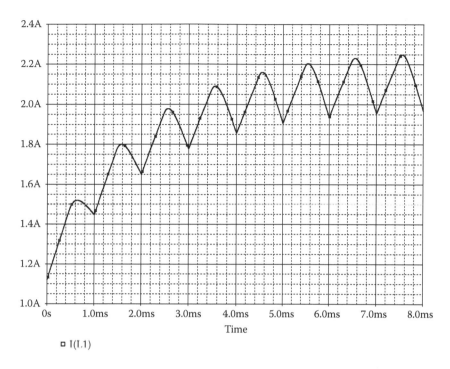

FIGURE 5.13
Transition process for the inductor current.

A useful function of **PSpice A/D** is **Initial Condition**. It enables the setting of the passive component's voltage or current values at the start of the analysis. In the case of the converter considered here, we can define the initial current of the inductor L1 and the voltage of the capacitor C1. The parameter **IC** for the capacitor should be set to 30 V, and for the inductor, IC = 1.98 A. These are the values obtained from the end of the last switching period of the previously done simulation.

The results of the analysis in the time domain in two periods of commutation are presented in Figure 5.15. They represent the steady state of the analyzed circuit and are the same as those obtained from **Mathematica**.

5.3.3 FFT Harmonics Analysis

In some cases there is a need to examine the spectrum of a time-dependent signal. As an example, let us analyze the load voltage of the converter presented earlier. It is possible to do so in two ways: directly in **Probe** and by setting the FFT parameters in the simulation profile.

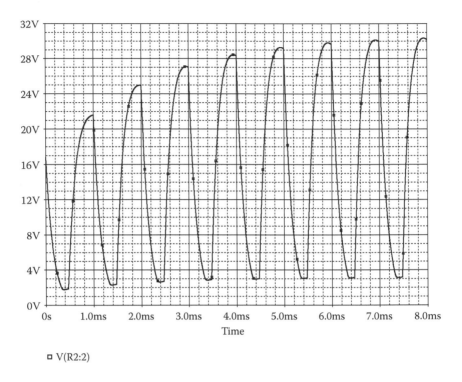

□ V(R2:2)

FIGURE 5.14
Transition process for the output voltage.

To convert the signal into its spectrum, we must first display it in the **Probe** window, as shown in Figure 5.16. It is recommended that the parameters of the transient simulation be set up for such a time interval that contains an integer number of periods. In our example, if the initial conditions are as in the last simulation, and the **Final Time** is equal to 2 ms, it means that, in two periods of the signal, we can observe the steady state of that voltage. To present the signal and its FFT as another plot, it is necessary to desynchronize them by choosing the option **UnsynchronizeXaxis** from the **Plot** menu. To transform the signal into its FFT, we use the **FFT** option from the **Trace** menu. It is necessary to set up the axes to a desired range.

The other method of performing the Fourier analysis is to define it in the simulation profile. Under the setup of **Transient Analysis**, there are fields to define the **Fourier Analysis**. If those fields are filled as follows:

- Enable Fourier = On
- Center Frequency = 1000 Hz
- Number of Harmonics = 5
- Output Var(s) = V(R2:2)

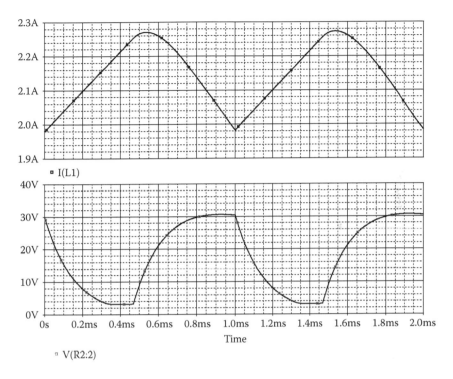

FIGURE 5.15
Steady state of the inductor current (upper window) and the load voltage (lower window).

then the output voltage V(R2:2) will be analyzed with the base frequency equal to 1 kHz, and the magnitudes and phases of the first five harmonics will be calculated. These results are saved at the end of the output file in the following form:

Fourier Components of Transient Response V($N_0002)
DC Component = 1.768876E + 01

Harmonic No.	Frequency (HZ)	Fourier Component	Normalized Component	Phase (Deg.)	Normalized Phase (Deg.)
1	1.000E+03	1.483E+01	1.000E+00	1.472E+02	0.000E+00
2	2.000E+03	1.028E+00	6.928E–02	3.660E+01	–2.579E+02
3	3.000E+03	1.966E+00	1.325E–01	1.281E+02	–3.136E+02
4	4.000E+03	4.279E–01	2.885E–02	4.253E+01	–5.464E+02
5	5.000E+03	7.057E–01	4.758E–02	1.273E+02	–6.089E+02

Total harmonic distortion = 1.638498E + 01 percent

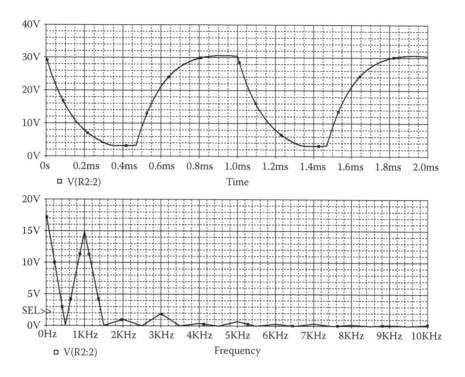

FIGURE 5.16
Waveform of the analyzed voltage (upper window) and its FFT diagram (lower window).

5.4 Processes in a System with Several Aliquant Frequencies

In Figure 5.17, a model of a DC/DC step-down converter, equivalent to that in Section 2.5, is shown. In those circuits, the switches **S2** and **S3** are in on/off states simultaneously. To avoid shortcircuits on the one hand and overvoltages on the other, they are controlled by the same pulse voltage source V3. Its actual parameters should be set as

- V1 = −10 V
- V2 = 10 V
- TD = 0 s
- TR = 1 ns
- TF = 1 ns
- PW = 80 μs
- PER = 100 μs

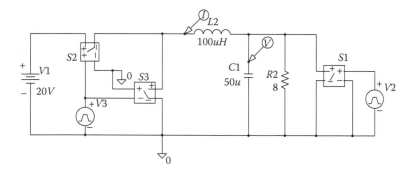

FIGURE 5.17
Circuit of the converter with the periodically commutated load.

The **S1**-voltage-controlled switch simulates the nonstationary load. Its resistance in the on state is equal to 8 Ω, and in the off state it is 1 MΩ. Parameters of the V2 source are as follows:

- V1 = 0 V
- V2 = 10 V
- TD = 0 s
- TR = 100 ns
- TF = 100 ns
- PW = 40 μs
- PER = 60 μs

To obtain the quasi-steady state of the circuit, as presented in Chapter 2, Section 2.5, the final time of the analysis is set to 6 ms. Results of the analysis are presented in Figures 5.18 (I_{L1} current) and 5.19 (the load voltage V(C1:1)).

The two final periods of those waveforms are shown. They are denoted as quasi-steady state because the frequencies of the pulses controlling the switches are aliquant. Therefore, all the waveforms are nonperiodical.

To avoid convergence problems, especially in time-domain analysis, it is recommended that properly high voltages be used in the control circuit. As voltages in the control and the main circuits are in the same ranges, there are fewer problems in calculating the algorithm to establish the precision of a calculation. Therefore, the **V2** and **V3** voltage values applied earlier are chosen as 10 V. The **Von** and **Voff** voltages of the switches **S** are assigned to 10 and −10 V.

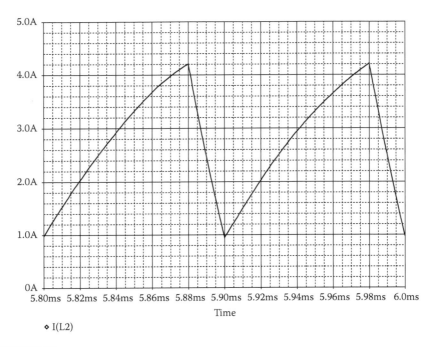

FIGURE 5.18
The quasi-steady state of the current.

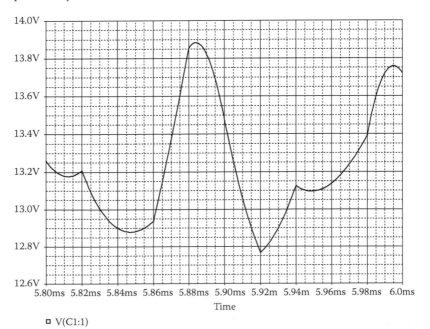

FIGURE 5.19
The quasi-steady state of the voltage.

5.5 Processes in Closed-Loop Systems

Let us consider the buck-boost DC/DC converter controlled with a voltage feedback. The model of the circuit is shown in Figure 5.20. The general problem with such a circuit is to transform the electrical values (voltages, currents, power, etc.) into control signals. The **ABM1** and **ABM2** blocks are used as the "amplifiers" in the voltage-control loop. Their input and output signals are refereed to 0 and can be considered as standard voltages (i.e., they can be connected to the other components). A relation between the output and input (or inputs) can be easily defined algebraically. For example, if the output signal must be negated and divided 100 times with respect to the input, the expression to be written is "$-(V(\%IN))/100$", where the " $(V(\%IN))$" notation means the input voltage. To define constant values, it is convenient to use the **Const** symbol and simply to define its value inside the **Value** field. We can also limit the signals to a desired range with the **Limit** part filling its low and high values.

The output voltage in this converter depends on a duty cycle of the signal controlling the switch **S1**. To obtain the proper duty factor in each period, the output voltage divided by -100 is compared with the 1.5 V reference signal. Next, that signal is amplified 1.6 times and is limited to between 0.5 and 4 V. This signal is compared with a ramp voltage and amplified 1000 times to obtain square pulses with the desired duty factor. These pulses, limited

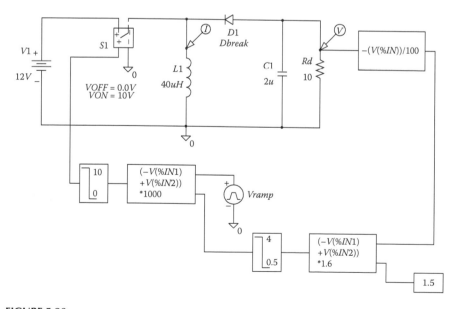

FIGURE 5.20
Buck-boost converter.

to the 0–10 V range, are used to control the **S1** switch. The parameters of the ramp voltage **Vramp** are

    DC = 0

    AC = 0

    V1 = 0

    V2 = 5 V

    TD = 0

    TR = 9998 ns

    TF = 1 ns

    PW = 1 ns

    PER = 10 μs

The parameters of the voltage-controlled switch **S1** are

    RON = 50 mΩ

    ROFF = 1 MΩ

    VON = 10 V

    VOFF = 0 V

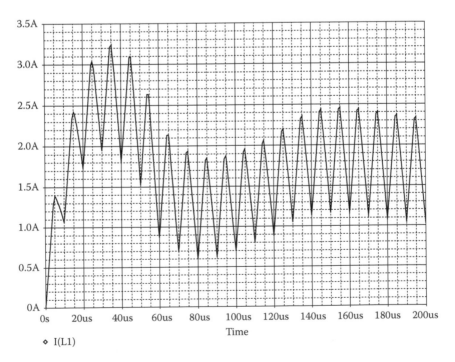

FIGURE 5.21
Transient process of the inductor current.

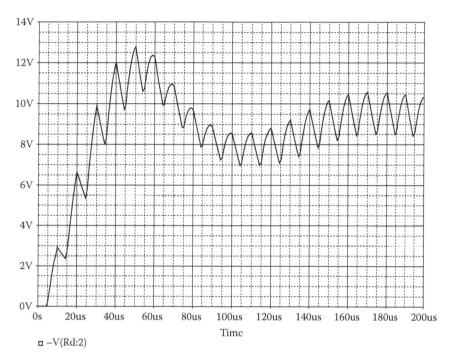

□ −V(Rd:2)

FIGURE 5.22
Transient process of the load voltage.

The transient processes of the inductor current I(L1) and reversed output voltage −V(Rd:2) are presented in Figures 5.21 and 5.22.

Changing the **Const** value, in our example is equal to 1.5 V; we can set up the output voltage to other values.

5.6 Modeling of Processes in Relay Systems

The main circuit of the relay system described in Chapter 3, Section 3.6, is shown in Figure 5.23. Components **S1-S4** and the **Vdc** voltage source represent a typical bridge VSI inverter. In such circuits, the pairs of switches **S1-S4** and **S2-S3** are in on or off states simultaneously. This is realized by the symmetrical setting of its parameters to **Von** = 10 V and **Voff** = −10 V values. The output filter is composed of **L1** and **C1** components. The resistor **R1** represents the load of the converter. Because the output voltage is bipolar (differential), it is convenient to transform it to the unipolar form (one pole should be connected with the **0** point) using the **E1** component. This component is the voltage-controlled voltage source with the parameter **Gain**

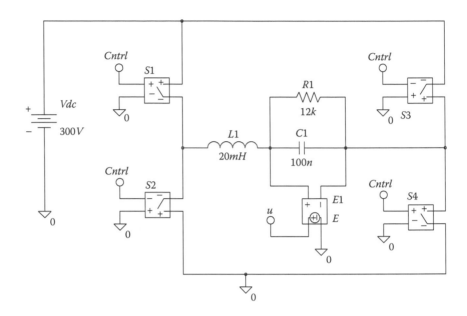

FIGURE 5.23
The main circuit of the DC/AC converter.

equal to 1. The one-pole signal **u** taken from **E1** is more convenient for the following control circuit.

The circuit shown in Figure 5.24 generates a rectangular control signal for the switches. Tracing of an input sinusoidal signal **Vgen** forms the alternating load voltage **u**. The load and generator voltages are compared by the **ABM2** block. Resistances of **R2** and **R3** determine the dead band of the Schmitt trigger. Next, in the **ABM1** block, the output signal is amplified

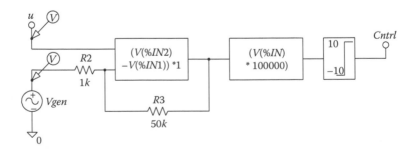

FIGURE 5.24
Control circuit of the converter.

100,000 times and, in the end, is limited to between –10 and 10 V by the **Limit** component.

The Schmitt trigger is a comparator with a positive feedback. It constitutes the hysteresis component, and its dead band depends on the gain defined by the R2/R3 ratio.

In our case, the input voltage Vgen parameters are

Vampl = 200 V

Freq = 10000 Hz

and all of the other values must be set as 0.

The results of transient analysis of the system for the first five periods, on time interval from 0 to 0.5 ms, are presented in Figure 5.25. The same analysis, but made for a longer time, is presented in Figure 5.26. As can be seen, the steady state of the system has been formed. The test results presented in the foregoing simulation are the same as those presented in Chapter 3, Section 3.6.

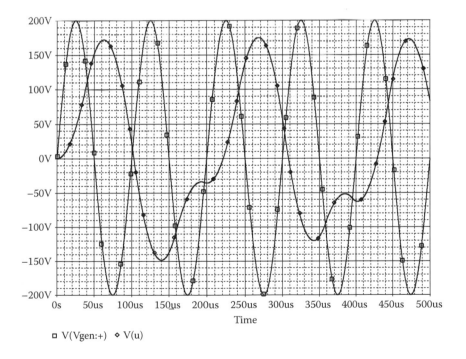

□ V(Vgen:+) ◇ V(u)

FIGURE 5.25
Transient process of the output voltage V(u) and generator voltage V(Vgen) for the generator frequency 10 kHz.

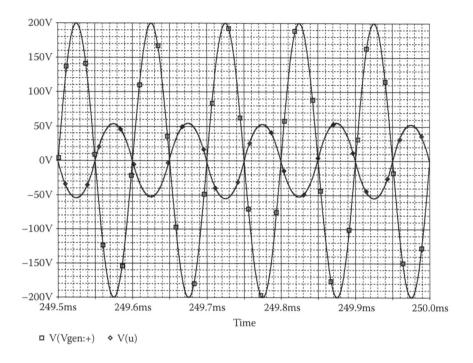

□ V(Vgen:+)   ◇ V(u)

FIGURE 5.26
Transient process of the output voltage V(u) and generator voltage V(Vgen) for generator frequency 10 kHz for the last 10 switches.

5.7 Modeling of Processes in AC/AC Converters

5.7.1 Direct Frequency Converter

A model of another kind of power electronics device is presented in Figure 5.27. This is a one-phase full-bridge direct AC/AC frequency converter. Pairs of switches S1, S3 and S2, S4 are in opposite states. When one pair is on, the second must be off. Moreover, duty factors of both control signals must be the same, equal to 1/2. The parameters of the switches are

ROFF = 1 MΩ
RON = 100 mΩ
VOFF = –9 V
VON = 9 V

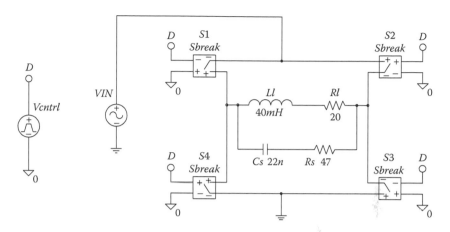

FIGURE 5.27
Direct frequency converter.

Bipolar control of the switches is achieved by reverse connection of pairs to the control source **Vcntrl**. The parameters of the source are

DC = 0 V

AC = 0 V

V1 = –10 V

V2 = 10 V

TD = 0 s

TR = 200 ns

TF = 200 ns

PW = 1.66645 ms

PER = 3.33333 ms

Since the rising and falling edges of the control voltage are equal to 200 ns, during those moments, none of the switches is really in an on state. Therefore, the load should be overvoltage protected by the Rs Cs snubber circuit. The converter is analyzed for transients with simulation of the parameters Final Time = 10 ms, Step Ceiling = 1 μs. The result of the analysis is presented in Figure 5.28. Because in such converters the number of control pulses must be equal to an even multiple of the AC supply half-period, the load voltage is rectangular with sinusoidal envelope.

5.7.2 Three-Phase Matrix-Reactance Converter

The three-phase matrix-reactance converter (MRC) is a kind of power electronics device that enables changing both the amplitude and frequency of

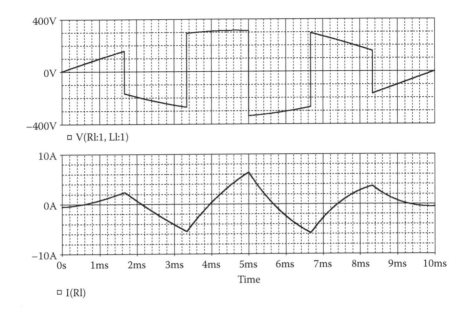

FIGURE 5.28
Load voltage and current of the converter.

the voltage. This particular device is based on two well-known topologies: the three-phase matrix converter, which enables connection of each input phase with each load terminal, and the buck-boost, for example, used in DC power suppliers. Its basic properties and the control strategy are described in Chapter 2, Section 2.7. The main circuit of the analyzed converter is presented in Figure 5.29. It consists of the following functional blocks: the three-phase AC power line with input filter (**VS1 ÷ 3, LS1 ÷ 3, CS1 ÷ 3**), the 3 × 3 matrix of bidirectional switches **SAA ÷ SCC**, boost inductors **LS1 ÷ 3**, load switches **SL1 ÷ 3**, and load resistors **RL1 ÷ 3** with filtering capacitors **CL1 ÷ 3**. The internal parameters of all voltage-controlled switches are

RON = 10 mΩ
ROFF = 100 kΩ
VON = 10 V
VOFF = 0 V

The amplitudes of supply voltages are VAMPL = 325 V.

The control method or modulation strategy of this converter is similar to standard three-phase matrix converters (MC). The only difference is that, in time intervals where, in a standard MC a zero vector is generated, in the MRC the load switches must be additionally turned on. In all other states of the circuit, those switches must be off. The proposed control circuit of the

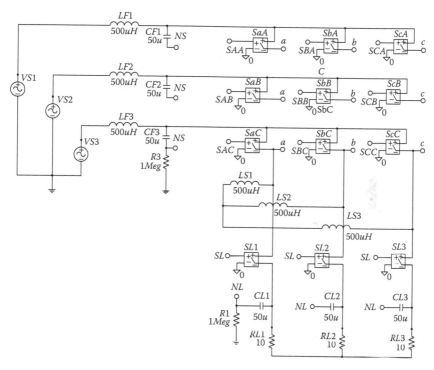

FIGURE 5.29
The main circuit of the matrix-reactance converter.

converter, based on the Venturini method, is presented in Figure 5.30. First, the input values are defined: time **t**, voltage transformation factor **q**, and a difference between the source and load angular frequencies ω_m. On the basis of the latter, three modulating signals are calculated; for example, in first phase, the formula written in the ABM block is

$$\text{VDF1} = (1+(2 \; q \; \cos (\omega_m \; \text{time}))/4.$$

The sums of the modulating signals are next compared with the ramp voltage Vramp, amplified and limited to (–1:11 V), thus generating PWM signals for the switches. The parameters of the ramp voltage source are

DC = 0 V

AC = 0 V

V1 = 0

V2 = 1 V

TD = 0 s

TR = 199 μs

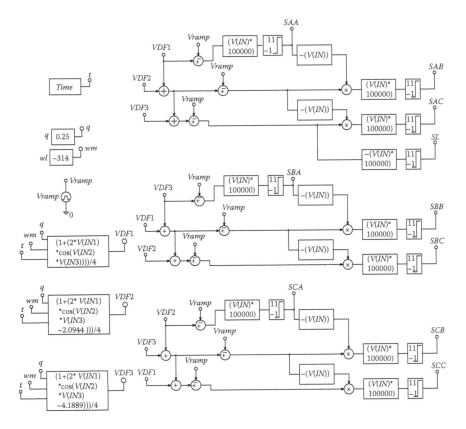

FIGURE 5.30
Control circuit of the matrix-reactance converter.

$$TF = 0.5 \ \mu s$$
$$PW = 0.5 \ \mu s$$
$$PER = 200 \ \mu s$$

The results of time-domain simulation are presented in Figure 5.31 (currents) and Figure 5.32 (voltages). To obtain the steady state of the circuit, the simulation results are printed from 15 ms. As can be seen, load values are smaller than in results presented in Chapter 2 because switching losses in the main circuit cause worse energetic efficiency.

5.7.3 Model of AC/AC Buck System

Let us consider a PSpice model of the AC line conditioner described in Chapter 4. Such devices are based on the well-known DC/DC buck topology, transformed into an alternating current by use of bidirectional switches in the

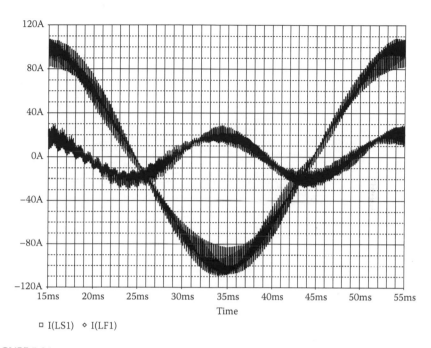

FIGURE 5.31
I_{LF1} and I_{LS1} inductor currents.

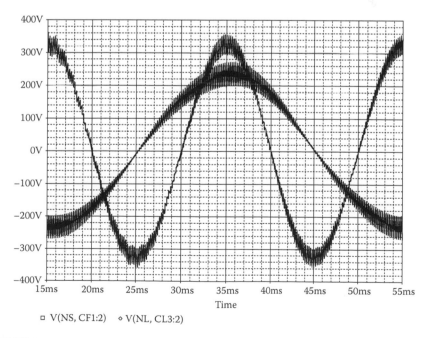

FIGURE 5.32
U_{CF1} and U_{CL1} capacitors voltages.

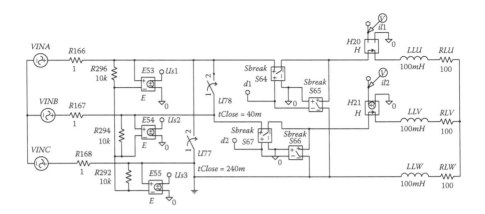

FIGURE 5.33
The main circuit of the AC/AC conversion system.

main circuit instead of transistors or diodes. Assuming that the commutation frequency is much higher than the input AC voltage, the load voltage depends linearly on the duty factors of the impulses that control the main switches. Because of the three-phase AC/AC three-wire connection, its structure can be simplified to the two-phase controller topology, as shown in Figure 5.33. The "C" phase of the circuit is a common wire, so the output currents and voltages could be regulated independently in phases "A" and "B." Sinusoidal voltage sources VINA-VINC with series R = 1 Ω resistors represent the three-phase power source. Its amplitude is equal to 325 V, and the frequency is 50 Hz. Components U77 and U88 are **Sw_Tclose** switches—in fact, they represent a time-dependent resistance and will be used to generate a line voltage imbalance. At the moment t = 40 ms, the U78 changes its resistance from 1 MΩ to 10 Ω. Star-connected voltage-controlled voltage sources E53–E55, with transform ratio **GAIN** = 0.03077, implement a separated line voltage measurement with nominal output amplitude equal to 10 V. It is necessary to connect them in parallel with three equal star-connected resistors. Moreover, to provide correctness of the voltage measure, the star point must not be grounded. S64–S67 voltage-controlled bidirectional switches realize the main function of the controller. As S64 and S67 are on, while S65 and S66 are off, the load is connected to the source. In the opposite state of these switches, the load is shorted, which enables continuity of the load current. Moreover, to avoid short circuits on the one hand and to minimize overvoltages on the other, the parameters VON and VOFF of the switches are set symmetrically. More precisely, in this circuit they are set as 9 V for on state and −9 V for off state for all switches. The control terminals of the switches connecting the load with the source, and the switches shorting the load, are just inversely connected. Such a configuration simplifies the control circuit because there is only one rectangular control signal needed for each phase of the converter, signed as

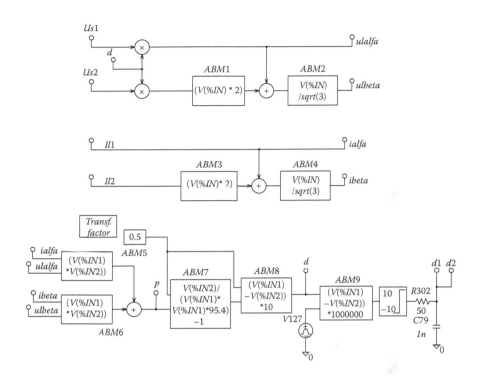

FIGURE 5.34
Control circuit of the AC/AC conversion system.

d1 and d2 in the scheme. They are measured, separated, and scaled by H20 and H21 current-controlled voltage sources with parameter GAIN = 3.225. The parameters of the models used in the main circuit are

- **VINA**: VAMPL = 325 V, FREQ = 50 Hz, PHASE = 0; for **VINB** and **VINC**, set phases equal to 120 and 240 degrees
- **E53 ÷ 55**: GAIN = 0.03077
- **U78**: tClose = 40 ms; ttran = 1 ms; Rclosed = 10 Ω, Ropen = 1 Meg
- **S64 ÷ 66**: ROFF = 30 kΩ, RON = 100 mΩ, VOFF = −9 V, VON = 9 V
- **H20, H21**: GAIN = 3.225

Control circuits of the converter are presented in Figure 5.34. Conceptually, the control is based on the instantaneous power theory. In two upper sub-circuits, the instantaneous values of the source voltage and load current are transformed into orthogonal α-β space-state vectors (in blocks **ABM1 ÷ 4**). Their coordinates can be observed as ulalfa–ulbeta and ilalfa–ilbeta signals, respectively. The ulalfa–ulbeta signals are obtained as a product of the source voltage and the calculated transformation factor **d**. It means that, in fact, those values represent the load voltage. This is a much easier way to obtain their waveforms because the real load voltage is chopped and would need conditioning (i.e., filtering or averaging). Expressions of main blocks and models in the control scheme are

- **ABM1** and **ABM3**: VIN*2
- **ABM2** and **ABM4**: VIN/sqrt(3)
- **ABM5** and **ABM6**: VIN1*VIN2
- **ABM7**: VIN2/(VIN1*VIN1*95.4)-1
- **ABM8**: (VIN1-VIN2)*10
- **ABM9**: (VIN1-VIN2)*1000 000
- **V127**: DC = 0 V, AC = 0 V, V1 = 0 V, V2 = 1 V, TD = 0 s, TR = 199.4 μs, TF = 200 ns, PW = 200 ns, PER = 0.2 ms
- initial conditions of the load current for LLU **IC** = −380 mA, for LLV **IC** = −1.26 A, for LLW **IC** = 1.65 A

Although the initial conditions are not necessary for inductor currents, they can be set up to shorten any transient states in the modeled circuit. Their values were obtained from previously done simulations.

As described in Chapter 4, the goal of the control method is to stabilize the value of the instantaneous power of the load. The power is determined as a sum of products of orthogonal currents and voltages (ABM5 and ABM6) and can be observed in point **d** in the lowest subcircuit in Figure 5.26. The next two blocks (ABM7 and ABM8) realize the closed-loop control system. The measured instantaneous power is being compared with the calculated one, and the control error is amplified 10 times. This operation determines the actual value of the voltage transformation ratio **d**. This **d** factor, compared in ABM9 with the ramp voltage, determines the duty factor of the signal controlling the main switches. Next, the rectangular signal is limited to −10 and 10 V. The last unit that forms the control signal is the RC delay circuit. This circuit forms the exponential shape of commutation processes as they occur in any real solid-state components.

5.7.4 Steady-State Time-Domain Analysis

Let us analyze the AC/AC converter previously described in the time domain. The parameters of the simulation are

Print Step = 100 μs
Final Time = 80 ms
Step Ceiling = 0.3 μs

Figure 5.35 presents the time waveform of the control signal **d**. This value determines the instantaneous voltage transformation factor of the conditioner being

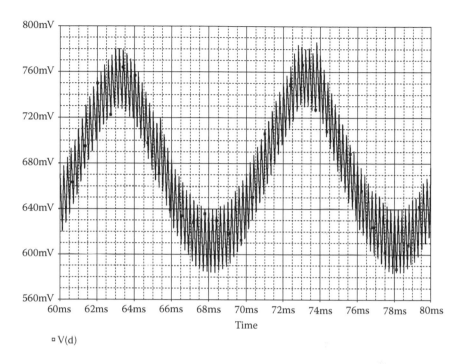

□ V(d)

FIGURE 5.35
Voltage transformation factor.

operated under voltage imbalance. As can be seen, the **d** factor is modulated by two components with frequencies equal to 100 Hz and 5 kHz. The first one provides the load voltage and current balance, and is generated by the control system. The second one is produced by the power conversion and the frequency of the ramp generator **V127**. Figure 5.36 presents waveforms of load currents in the steady state. As can be seen, their amplitudes are equal, a fact confirming the correctness of the presented model and its control method.

5.8 Static Characteristics of the Noncompensated DC Motor

In this part the PSpice model of the DC motor described in Section 4.2 will be shown. The proposed electrical circuit realization of Equations 4.32 and 4.33 is shown in Figure 5.37. The top right circuit is a realization of the excitation part of the motor. It consists of the V_E and R_E components, which represent the excitation voltage and resistance, respectively. The current-controlled voltage source **H_TORQ** implements a relation between the armature current and the weakening of the magnetic field. Its Gain = 0.755 results from

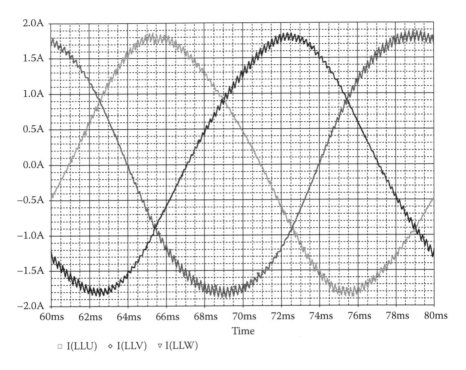

FIGURE 5.36
Time waveforms of load currents.

the slope coefficient of the theoretical line shown in Figure 4.7. The **H_KPHI** source, with **Gain** = –1 provides the straight calculation of the $k\Phi$ parameter of the machine. The top left circuit represents the armature. RA and LA are the armature resistance and inductance, respectively. The voltage-controlled voltage source **E_EMF**, with **Gain** = 1, represents the back electromotive force of the machine. Its value is proportional to the product of the actual value of $k\Phi$ and the angular velocity of the rotor ω. The lower circuit is the electrical representation of a mechanical part of the motor. An electromechanical torque is equal to the product of the $k\Phi$ and the armature current. In the model, its value is represented in volts. At the end of the circuit, there is a load torque, simulated by the voltage source **V_ML**. The series inductance L_J represents a moment of inertia, and the resistance R_F simulates mechanical losses. The angular velocity is expressed in amperes. The current-controlled voltage source H_W changes the signal into the voltage value **w**. The **ABM** block allows recalculation of the angular speed ω into the rotation speed **n** [1/min] by simply multiplying the value by 30 and dividing it by π.

Let us examine how that circuit can show relations between the load torque and weakening of the magnetic field or armature current. For the determination of any static characteristics of the machine, a DC Sweep analysis seems to be the most proper. We set the armature voltage V_A = 420 V and the excitation

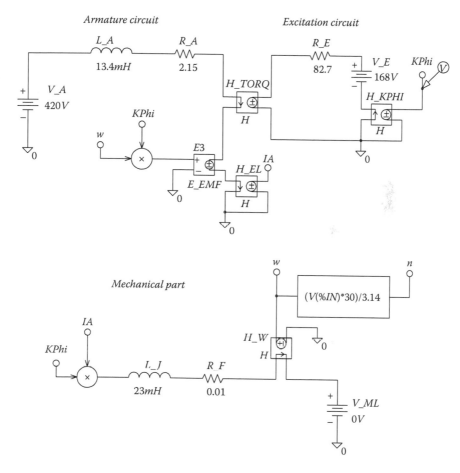

FIGURE 5.37
PSpice model of the noncompensated DC Motor.

voltage V_E = 210 V, both as nominal values. The load torque voltage **V_ML** will be the swept value, so, in the scheme, its value is optional. In fact, the value of **V_ML** is set in the simulation profile. The parameters of the analysis are

Sweep type: linear
Swept var type: V_ML
Start Value = 0 V
End Value = 200 V
Increment = 0.1

As a result, the load torque of the machine increases linearly, and any responses of the circuit to it are calculated.

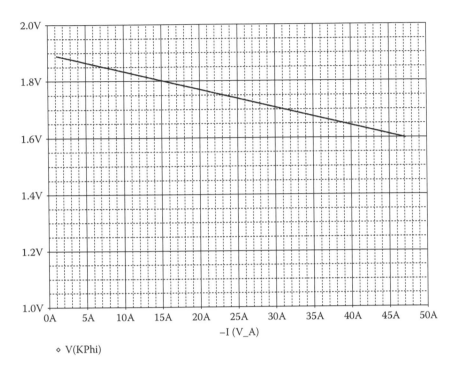

FIGURE 5.38
Field weakening for nominal U_A and U_E voltages versus the armature current.

Let us examine the model for the influence of the armature current on field weakening. The value of the armature current mainly depends on the load torque M_L. In our model, it is simulated by the **V_ML** voltage, where 1 V corresponds to 1 Nm of the load torque. We can easily plot the curve representing field weakening versus armature current $k\Phi = f(IA)$, as shown in Figure 5.38. It is obtained by choosing the **V(kPhi)** plot, setting as the x axis variable the armature current **–I(V_A)** and scaling the axes.

To examine the dependence between the armature current and the load torque $I_A = f(M_L)$, we can use the same model and simulation. Excitation and armature voltages are nominal, and the load should increase. The plot of the curve is shown in Figure 5.39. As can be seen, the armature current does not grow linearly with the load torque. This is caused by power losses in the motor represented in our model by the resistor **R_A** and the earlier-mentioned field weakening. The resistor **R_F** represents friction losses in the motor and its value, the idle current of the machine. Its value cannot be 0 because of a time constant of the mechanical circuit that equals $\tau = \frac{L\_J}{R\_F}$.

Insofar as in our model the torque depends on the magnetic flux, we can also examine the phenomenon of curving of the torque and determine its critical value. The plot of the torque versus the armature current is presented in Figure 5.40. This is obtained by use of the same DC sweep simulation.

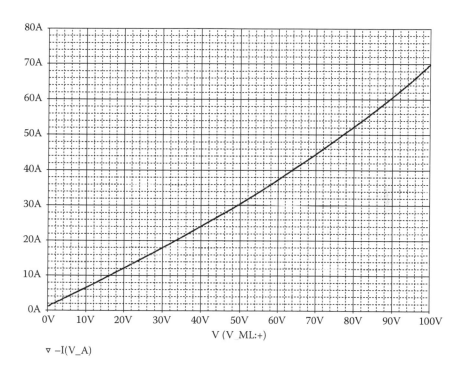

$\triangledown$ $-I(V\_A)$

FIGURE 5.39
Armature current as a function of the load torque for the nominal supply.

As the x-axis variable, the armature current $- \mathbf{I(V\_A)}$ is chosen, and as the observed value, the torque $\mathbf{V(V\_ML{:}+)}$ is chosen. As can be seen, the critical torque equals 112 Nm, and the same value is presented by the mathematical model in Chapter 4, Section 4.2. The only difference between Mathematica and PSpice simulations is that, in this simulation, the input value is the torque. In fact, the dependence is calculated reversely as $I_A = f\,(M_L)$. As a result, the plot of the torque over the critical value does not decrease since the function can have only one value for one argument. Let us try to avoid that problem during the examination of the mechanical characteristics of the motor. As we want to observe the torque versus rotation, the input value must be a current force. We have to change the voltage source $\mathbf{V\_ML}$ into the current source $\mathbf{I\_ML}$ and set the DC sweep analysis parameters as

Sweep type: linear
Swept var type: I_ML
Start Value = −50 A
End Value = 250 A
Increment = 0.1

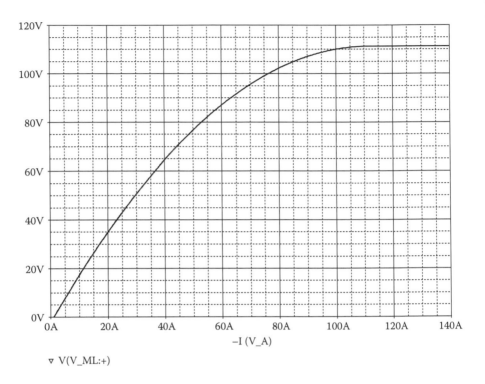

FIGURE 5.40
Curving of the torque characteristics.

All the remaining parameters should be left as in the previous simulation. In Figure 5.41, the V(I1:+) versus rotation speed is shown, which is the M/n characteristic of the model.

This is the particular curve that shows that the mechanical characteristics of noncompensated and asynchronous motors are similar. The same characteristic is presented in Chapter 4, Figure 4.11.

5.9 Simulation of the Electrical Drive with Noncompensated DC Motor

The model of the control system of a DC drive for the earlier-presented noncompensated motor is shown in Figure 5.42. The parameters of the motor are the same as presented in Section 5.7 and Figure 5.37, but in this case, the load torque is dependent on the square of the motor speed. The value of the load is modeled by the **ABM_ML** block, which recalculates the angular speed

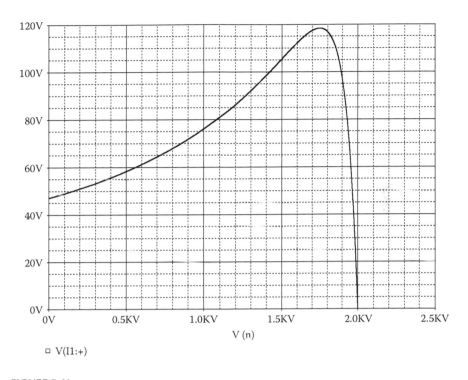

FIGURE 5.41
The mechanical characteristics of the model.

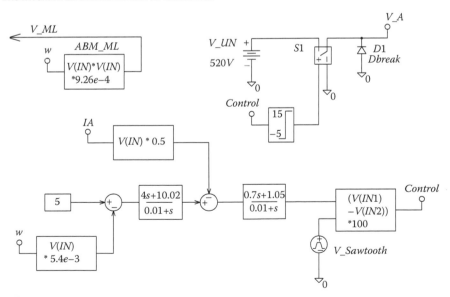

FIGURE 5.42
The model of the speed and current controller of the DC drive.

into the **V_ML** value. Components **V_UN**, **S1**, **D1**, and the internal inductance of the motor **L_A** simulate the Buck converter that allows control of the
armature voltage V_A between 0 and 520 V. Parameters of the switch **S1** are

ROFF = 1 Meg
RON = 10 m
VOFF = 0 V
VON = 10 V

The reference $\omega = 5$, set in the constant block, is compared with the actual
speed of the motor and regulated next by the PI controller. The formula in
the block of the speed controller is $v_{out} = \frac{4s+10.02}{0.01+s} v_{in}$ and it differs from the
mathematical model by the 0.01 constant in the denominator. Its value is neg-
ligible for regulation processes, but it is necessary to put it there because of
possible convergence problems during the simulation. Similar to the speed,
the prescribed current is compared with an actual value and regulated by
the PI block by the formula $v_{out} = \frac{0.7s+1.05}{0.01+s} v_{in}$. Finally, the output signal forms

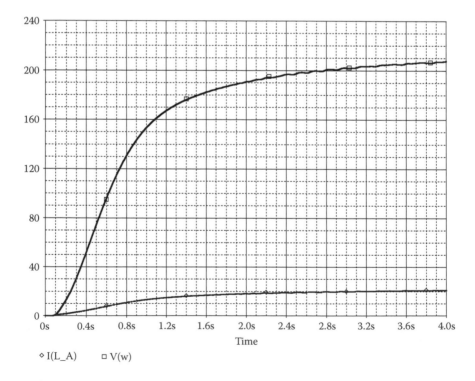

FIGURE 5.43
Motor starting: armature current I(I_LA) and rotor speed V(w) versus time.

a duty factor of the control pulses by comparison with the **V_Sawtooth** voltage whose internal parameters are as follows:

DC $= 0$

AC $= 0$

V1 $= 0$

V2 $= 5$

TD $= 5$ m

TR $= 50$ n

TF $= 99.9$ u

PER $= 100$ u

At the end of the control system is the voltage comparator with gain $= 100$ and limit block, which finally forms the control impulses for the main switch **S1**.

To observe the transient states during the start of the drive, the model is simulated in the time domain for 4 s. Due to a great number of commutations and the long time of the analysis, it is recommended that interested values in the schematics be marked, and limit the data to be collected to the marked ones. The results of the simulation are presented in Figure 5.43.

One can see that the results obtained are similar to those presented in Chapter 4, Section 4.2.

References

1. Bellman R. (1976). *Introduction to Matrix Analysis*. Nauka, Moscow, USSR, 352 pp. [in Russian]
2. Bromberg P. V. (1967). *Matrix Methods in the Theory of Relay and Impulse Control*. Moscow, Nauka, USSR, 324 pp. [in Russian]
3. Davies P. I., Higham N. J. (2005). *Computing f(A)b for Matrix Function f*. School of Mathematics, University of Manchester, Manchester, M199PL, p. 11.
4. Director S., Rorer R. (1974). *Introduction to System Theory*. Mir, Moscow, 464 pp. [in Russian]
5. Gantmacher F. R. (1977). *The Theory of Matrices*. Chelsea, New York.
6. Kasperek R. (2003). Control algorithms of the PWM AC line conditioners under unbalanced input voltage, APEDIA conf. mat., Tallin.
7. Korotyeyev I. Ye. (1999). The calculation of steady-state processes in circuits of voltage converters, which are working on periodical load. Modeling and Simulation of Electric Machines, Converters and Systems—Electrimacs '99. *Proceedings of the 6th International Conference*, Lisboa, Portugal, 1999. Vol. 3, pp. 215–220.
8. Korotyeyev I. Ye., Fedyczak Z. (1999). Analysis of steady-state behaviour in converters with changed topology. Supply System of Electrotechnical Devices and Systems, Kiev, Ukraine. *Technical Electrodynamics*, No. 1, pp. 31–34. [in Russian]
9. Korotyeyev I. Ye., Klytta M. (2002). Stability analysis of DC/DC converters. Power Electronics and Energy Efficiency, Kiev, Ukraine. *Technical Electrodynamics*, No. 1, pp. 51–54.
10. Korotyeyev I. Ye., Fedyczak Z. (2002). Calculation of transient behaviours in AC converters. Power Electronics and Energy Efficiency, Kiev, Ukraine. *Technical Electrodynamics*, No. 1, pp. 43–46. [in Russian]
11. Korotyeyev I. Ye. (2003a). Analysis of periodic, quasiperiodic and chaotic processes in tracing systems. Power Electronics and Energy Efficiency, Kiev, Ukraine. Technical Electrodynamics, No. 1, pp. 67–72. [in Russian]
12. Korotyeyev I. Ye. (2003b). Analysis of periodic and chaotic processes in inverter under tolerance band control. 3rd International Workshop on Compatibility in Power Electronics—CPE 2003. Polska, Gdańsk-Zielona Góra, Poland, pp. 298–303. [CD-ROM]
13. Korotyeyev I. Y., Kasperek R. (2004a). Three-phase AC conditioner with instantaneous power control: Mathematical modelling of processes. Problems of Present-Day Electrotechnics, Kiev, Ukraine. *Technical Electrodynamics*, No. 5, pp. 99–102. [in Russian]
14. Korotyeyev I. Y., Kasperek R. (2004b). Three-phase AC conditioner with instantaneous power control: Stability analysis and processes modelling. Problems of Present-Day Electrotechnics, Kiev, Ukraine. *Technical Electrodynamics*, No. 5, pp. 95–98. [in Russian]

15. Korotyeyev I. Ye. (2004). Stability calculation of DC converter for small switching period of power switches. Power Electronics and Energy Efficiency, Kiev, Ukraine. *Technical Electrodynamics*, No. 3, pp. 114–117. [in Russian]

16. Korotyeyev I. Ye., Klytta M. (2005). Properties and characteristics of non-compensated DC motors. Power Electronics and Energy Efficiency, Kiev, Ukraine. *Technical Electrodynamics*, No. 3, pp. 26–27.

17. Korotyeyev I. Ye., Klytta M. (2006a). Real properties of non-compensated DC motors. Problems of Present-Day Electrotechnics, Kiev, Ukraine. *Technical Electrodynamics*, No. 7, pp. 31–34.

18. Korotyeyev I. Ye., Klytta M. (2006b). Stating characteristics of electrical drive with non-compensated DC motor. Power Electronics and Energy Efficiency, Kiev, Ukraine. *Technical Electrodynamics*, No. 5, pp. 38–41.

19. Korotyeyev I. Ye., Fedyczak Z. (2008a). Analysis of steady-state processes in matrix converter. Problems of Present-Day Electrotechnics, Kiev, Ukraine. *Technical Electrodynamics*, No. 1, pp. 91–96. [in Russian]

20. Korotyeyev I. Ye., Fedyczak Z. (2008b). Analysis of transient and steady-state processes in three-phase symmetric matrix-reactance converter system. Power Electronics and Energy Efficiency, Kiev, Ukraine. *Technical Electrodynamics*, No. 2, pp. 104–109.

21. Microsim PSpice Design Lab User's Guide.

22. Middlebrook R. D., Ćuk S. (1976). A general unified approach to modeling switching converter power stages. *IEEE Power Electronics Specialists Conference Record*, PESC' 76, Cleveland, OH, pp. 18–34.

23. Muhammad H. Rashid, Hasan M. Rashid. (2005). *SPICE for Power Electronics and Electric Power*, 2nd Ed., CRC Press, Boca Raton, FL.

24. Ned Mohan, Tore M. Undeland, William P. Robbins. (2002). *Power Electronics: Converters, Applications, and Design, Media Enhanced—with CD* (3rd Ed.), John Wiley & Sons, New York.

25. Pupkov K. A., Kapalin V. I., Jushchenko A. S. (1976). *Functional Series in Theory of Non-linear Systems*, Nauka, Moscow, USSR. [in Russian]

26. Rozenwasser N. Ye., Yusupov R. M. (1981). *Sensitivity of Automatic Control Systems*. Nauka, Moscow, USSR, 464 pp. [in Russian]

27. Rudenko V. S., Zhuykov V. Ya., Korotyeyev I. Ye. (1980). *Calculation of Devices of Industrial Electronics*. Technics, Kiev, Ukraine, 135 pp. [in Russian]

28. Strzelecky R., Korotyeyev I. Ye., Zhuykov V. Ya. (2001). *Chaotic Processes in Systems of Power Electronics*. Avers, Kiev, Ukraine, 197 pp. [in Russian]

29. Tolstoy G. P. (1951): *Fourier Series*, Gos. Izd. Techn.-teor. Lit., Moscow, USSR. [in Russian]

30. Tsypkin Ya. Z. (1974). *Relay Control Systems*. Nauka, Moscow, USSR, 576 pp. [in Russian]

31. Venturini M., Alesina A. (1980). The generalized transformer: A new bi-directional sinusoidal waveform frequency converter with continuously adjustable input power factor, *IEEE Power Electronics Specialists Conference Record*, PESC'80, Atlanta, GA, pp. 242–252.

32. Veszpremi K., Hunyar M. (2000). *New Application Fields of the PWM IGBT AC Chopper*, IEEE PEVD Conference Publication, No. 475, London, pp. 46–51.

33. Waidelich D. L. (1946). The steady-state operational calculus. *Proceedings of the the Institute of Radio Engineers* (IRE), IRE/IEEE.

34. Zhuykov V. Ya., Korotyeyev I. Ye., Ryabenky V. M., Pavlov G.V., Racek V., Vegg A., Liptak N. A. (1989). *Closed-up Systems of Electrical Power Transform.* Technics, Kiev, Ukraine, Alpha, Bratislava, Slovakia, 320 pp. [in Russian]

35. Zhuykov V. Ya., Korotyeyev I. Ye. (2000). Conditions of existence of strange attractor for PWM Systems. Problems of Present-Day Electrotechnics, Kiev, Ukraine. *Technical Electrodynamics*, No. 1, pp. 64–68. [in Russian]

Index